THE FUTURE
WE MAKE

Maker Faire® MAY 17–19
BAY AREA SAN MATEO EVENT CENTER
makerfaire.com

CONTENTS

Make: **Volume 68** April/May 2019

ON THE COVER:
A new crop of dev boards provide options for makers. Clockwise from left: Nvidia Jetson AGX Xavier, Asus Tinker Board S, Raspberry Pi 3 Model A+, Alchitry Cu, Adafruit Circuit Playground Express, Arduino MKR Vidor 4000, Particle Xenon, Adafruit Feather M4 Express, Adafruit Metro M4 Express, Google Coral Dev Board.

Photo: Mark Madeo

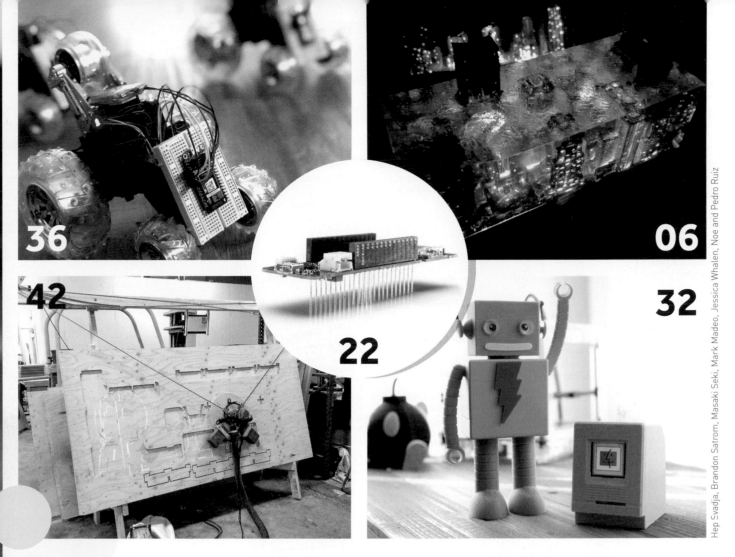

36

06

42

22

32

Hep Svadja, Brandon Satrom, Masaki Seki, Mark Madeo, Jessica Whalen, Noe and Pedro Ruiz

Make:

> "To achieve great things, two things are needed: a plan and not quite enough time." —Leonard Bernstein

EXECUTIVE CHAIRMAN & CEO
Dale Dougherty
dale@makermedia.com

CFO & COO
Todd Sotkiewicz
todd@makermedia.com

EDITORIAL

EXECUTIVE EDITOR
Mike Senese
mike@makermedia.com

SENIOR EDITORS
Keith Hammond
khammond@makermedia.com
Caleb Kraft
caleb@makermedia.com

EDITOR
Laurie Barton

PRODUCTION MANAGER
Craig Couden

MAKE: LEARN EDITOR
Roger Stewart
roger@makermedia.com

BOOKS EDITOR
Patrick Di Justo

CONTRIBUTING EDITORS
William Gurstelle
Charles Platt

CONTRIBUTING WRITERS
Massimo Banzi, Phil Bowie, Sam Brown, Michael Carroll, Jeremy Cook, Igor Daemen, Andy Forest, Sam Freeman, Limor Fried, Eric Hagan, Patrick Kinnamon, Adhith Mathradikkal, Angela Melick, Forrest M. Mims III, Dario Pennisi, David Romano, Noe Ruiz, Pedro Ruiz, Billy Rutledge, Brandon Satrom, Chris Skiles, Bar Smith, Zach Supalla, Marius Taciuc, Luke Valenty, Xo Ne Un

DESIGN, PHOTOGRAPHY & VIDEO

ART DIRECTOR
Juliann Brown

SENIOR VIDEO PRODUCER
Tyler Winegarner

MAKEZINE.COM

WEB/PRODUCT DEVELOPMENT
Rio Roth-Barreiro
Maya Gorton
Bill Olson
Pravisti Shrestha
Travis Stone
Alicia Williams

CONTRIBUTING ARTISTS
Eric Hagan, Mark Madeo, Angela Melick, Hep Svadja

ONLINE CONTRIBUTORS
Gareth Branwyn, Chiara Cecchini, Gretchen Giles, Adora Svitak

PARTNERSHIPS & ADVERTISING
makermedia.com/contact-sales or partnerships@makezine.com

SENIOR DIRECTOR OF PARTNERSHIPS & PROGRAMS
Katie D. Kunde

DIRECTOR OF PARTNERSHIPS
Shaun Beall

STRATEGIC PARTNERSHIPS
Cecily Benzon
Brigitte Mullin

DIRECTOR OF MEDIA OPERATIONS
Mara Lincoln

DIGITAL PRODUCT STRATEGY

SENIOR DIRECTOR, CONSUMER EXPERIENCE
Clair Whitmer

DIGITAL COMMUNITY PRODUCT MANAGER
Matthew A. Dalton

MAKER FAIRE

MANAGING DIRECTOR
Sabrina Merlo

COMMERCE

PRODUCT MARKETING MANAGER
Ian Wang

OPERATIONS MANAGER
Rob Bullington

PUBLISHED BY
MAKER MEDIA, INC.
Dale Dougherty

Copyright © 2019 Maker Media, Inc. All rights reserved. Reproduction without permission is prohibited. Printed in the USA by Schumann Printers, Inc.

Comments may be sent to:
editor@makezine.com

Visit us online:
make.co

Follow us:
🐦 @make @makerfaire @makershed
google.com/+make
f makemagazine
makemagazine
▶ makemagazine
twitch.tv/make
makemagazine

Manage your account online, including change of address:
makezine.com/account
866-289-8847 toll-free in U.S. and Canada
818-487-2037,
5 a.m.–5 p.m., PST
cs@readerservices.makezine.com

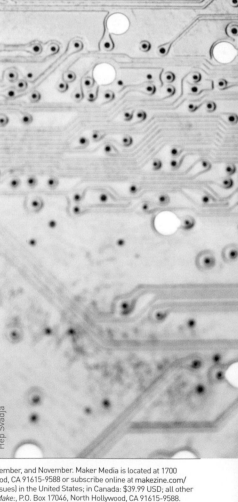
Hep Svadja

CONTRIBUTORS

What are you working on next?

Adhith Mathradikkal
Kozhikode, Kerala, India
(Sound Blink)
I'm very excited about my next project, which is a scaled down automatic bowling alley. I'm working on having it sort and distribute the knocked out pins, display scores, and return the ball back to the player.

Sam Freeman
Sonoma, CA
(Reproducible Ring)
I keep trying to get back to prop projects, but first I have a pair of wedding bands to build.

Juliann Brown
Oakland, CA
(Glass Block Night Light)
I'm trying to figure out how to not burn my arm every time I take something out of the oven, as well as honing my soldering skills.

Michael Carroll
Chalfont, PA
(Scrappy Circuits)
I'm finishing up a collection of some very unique MakeyMakey video game controllers. As a break, I started to make a record player from items at the dollar store.

Issue No. 68, April/May 2019. *Make:* (ISSN 1556-2336) is published bimonthly by Maker Media, Inc. in the months of January, March, May, July, September, and November. Maker Media is located at 1700 Montgomery Street, Suite 240, San Francisco, CA 94111. SUBSCRIPTIONS: Send all subscription requests to *Make:*, P.O. Box 17046, North Hollywood, CA 91615-9588 or subscribe online at makezine.com/offer or via phone at (866) 289-8847 (U.S. and Canada); all other countries call (818) 487-2037. Subscriptions are available for $34.99 for 1 year (6 issues) in the United States; in Canada: $39.99 USD; all other countries: $50.09 USD. Periodicals Postage Paid at San Francisco, CA, and at additional mailing offices. POSTMASTER: Send address changes to *Make:*, P.O. Box 17046, North Hollywood, CA 91615-9588. Canada Post Publications Mail Agreement Number 41129568. CANADA POSTMASTER: Send address changes to: Maker Media, PO Box 456, Niagara Falls, ON L2E 6V2

PRINTED WITH SOY INK

Nay for Quotes
and
Yay for SLA

MAKE IT OFFICIAL

In "Brite Done Rite" (Vol. 66, page 12), Gareth Branwyn reports Jason Webb's job title as "Creative Technologist." I think *Make:* magazine should take the bull by the horns and make it official: remove the quotes and make the job title Creative Technologist a thing. Berry College offers a Creative Technologies undergraduate major. University of Colorado Boulder offers a master's degree in Creative Technology and Design. The job title has been used in recent job listings for Apple, Mercedes-Benz, Georgia-Pacific, Oracle, Nordstrom, and Stanford University. So, I think the quotes are no longer needed. Creative Technologist is a job title for companies who want to hire makers but want a slightly fancier title.

–*John Grout, via email*

A MOST EXCELLENT ISSUE

Congratulations to you and all your staff on Vol. 66. I loved everything starting with your fun and unrepentant "Prison Ban of the Month" ("Reader Input," page 5) and through the great article by fellow Penn alum Nisan Lerea ("Under Pressure," page 16) and beyond. The whole thing! Fantastic tip for me by Peter Tas on laser kerf cut bending ("Laser Cutters," page 32). And I loved the leather-forming project, too ("3DP Leather Press," page 62). I felt the magazine this month was prepared just for me! To top it off, you indulged me with my pet peeve against patent proselytizing — with nary a tiny toot of any IP horn from cover-to-cover.

–*Arthur Krolman, via email*

MORE SLA REVIEWS, PLEASE

I look forward to the 3D printer edition every year. I have used your magazine to reference what's new and reviews to broaden my knowledge of digital fabrication. I have two FFF 3D printers (a delta and Taz 4) that I used for making parts for pinball machines.

I have always been interested in SLA printers but couldn't justify the cost. With your latest issue 66 showing just one SLA I was disappointed. Seeing the price point at $600 sparked my interest and started an internet search to find the price has dropped to less than $450. I pulled the trigger and purchased an Anycubic Photon SLA. Please continue to include SLA printers in your reviews.

–*Kenneth FitzHenry, via email*

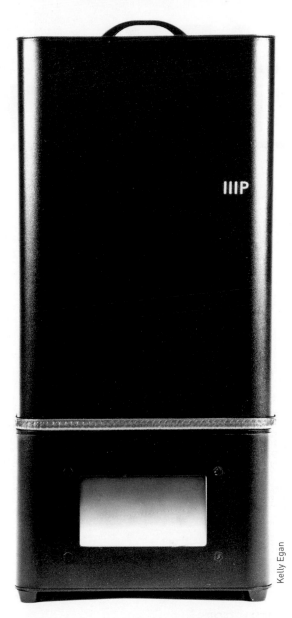

Kelly Egan

Christian "Chonky Lynx" Restifo
@restifo

Follow

When I was 7, I would thumb through the Sears catalog and mark pages for my parents for Christmas gift ideas. Maybe I'll do this with @make at 47 with my wife.

3:40 PM - 4 Dec 2018

MADE ON EARTH

Backyard builds from around the globe

Know a project that would be perfect for Made
on Earth? Let us know: *editor@makezine.com*

WATER WORLDS

INSTAGRAM.COM/MASAKI1006

A submerged cityscape, crumbling to become the debris of the ocean floor, is common imagery in dystopian media. There's just something about the interplay of light through water, familiar places, and urban decay that fascinates us. Japanese artist **Masaki Seki**, like many, was drawn to this imagery, but his results have been anything but common.

Seki's submerged dioramas depict flooded cities, crumbling amusement parks, and ghostly lit skyscrapers with immaculate detail. There is nearly no end to the minutiae of these dioramas; the closer you look, the more elements seem to pop out at you.

One aspect that seems to help add to the attraction is how the dioramas appear to be defying physics, a perfect slice of ocean standing without support. This is because after painstakingly sculpting, carving, and painting the environments, Seki pours resin for the water, which he says is one of the most difficult parts — so as to not introduce bubbles. After that are the finishing touches, and his magical pieces come to life. —*Caleb Kraft*

Masaki Seki

BEARLY THERE

DEBBIELAWSON.COM

Have you ever noticed how some inanimate objects happen to look like a face, such as faucets with long dripping noses, or power outlets staring surprised at the room you're in? This creative pattern recognition has a name, *pareidolia*.

When London-based artist **Debbie Lawson** flexes her pareidolia muscles, she goes in a slightly different direction, letting it drive her fascinating sculptures.

"Images always seem to leap out at me from patterns and textures, and once I've spotted them, they are impossible to 'un-see,'" Lawson says. "I have the same issue with furniture — it often suggests a particular animal or human trait to me. Once this idea sticks in my head, I work with it to bring that animal/trait to the foreground."

This series of works, which brings bears to life from the patterns in oriental rugs, was inspired by Lawson's travels around the world.

"I became entranced by the history and geographical spread of oriental rugs, finding them in the former Russian states where they have been handmade for centuries, as well as parts of Europe," she says.

Starting from a set of two identical rugs, Lawson sculpts the bear from scratch. Chicken wire, tape, and tons of effort give the bear its structure, then one of the rugs is sacrificed to create the "skin." The other rug remains intact to act as the backdrop.

The final result leaves you wondering at a glance, "did I really see that? Or was it just my imagination acting up?" — it may just be a clear peek through Lawson's mind's eye. —*Caleb Kraft*

Mike Warren

QUEST FOR FIRE MICHAELSAURUS.COM

Skateboarding has always been an unsafe hobby. You're hurling yourself along at running speed on a platform that is, by design, unstable. This is further punctuated by the fact that most skateboard tricks involve variations of jumping and flipping the board around in ways that seem nearly impossible to land.

Autodesk designer **Mike Warren** decided to take that unsafe element of skateboarding and ramp it way, way up, so he strapped what amounts to a rudimentary flamethrower to the bottom of his board. Now, as he scoots along, he can press a button with his toe, releasing a liquid fuel below the board that is ignited by a barbecue grill lighter.

"I made this project because I love mashing up ideas to make crazy things," Warren says. "This project went from 'I wonder if I could make it work' to 'how much fire is too much?' The answer is never enough."

In another twist, the sport is something Warren doesn't even

participate in. "This is going to sound crazy, but I don't know how to ride a skateboard," he says. "I had a vision of something that hadn't been made before and got to designing. Though I love the finished product, a lot of the satisfaction I get from making is solving the problems along the way. It's the journey, not the destination. Though, the destination of lighting up roadways and scaring the locals is pretty amazing."

The results of his skateboard are stunning. Like a scene from *Back To The Future*, Warren's board leaves flaming tracks in its wake as he shoots past, winning the admiration of daredevils everywhere despite the impracticality and inherent danger — so much so that his Instructable for it was removed. "I got in a little trouble for it at work since fire=danger, but secretly I love making polarizing projects that get people excited/nervous. I feel like if you aren't getting people talking about your project one way or another, then you're not doing it right." —*Caleb Kraft*

Make: membership

Becoming a Make: member means being part of a worldwide community of creative individuals who — like you — are inspired by raw ingenuity and passionate about exploring new project ideas, collaborating with others, and learning new hands-on skills. Each Make: member gains access to valuable member-only benefits, including:

- **Discounts** on Maker Faire tickets and online purchases through Maker Shed

- **Pre-notification** and **early registration** for online skill-building workshops and live events

- **Exclusive access** to the complete collection of Make: magazine digital volumes and video footage featuring top makers such as Laura Kampf and Massimo Banzi

Help us shape how people live, learn, and earn.

Become a Make: member today make.co

OUTSIZED
UPCYCLING

The Monterey Bay Aquarium teamed up with artists to create a life-sized blue whale out of recycled single-use plastic — to raise public awareness about how much of it washes into the world's oceans

Written by
Keith Hammond

THE BIGGEST BEAST THAT EVER LIVED, THE BLUE WHALE DWARFS EVEN THE DINOSAURS: 150 TONS, IF YOU COULD WEIGH ONE. Sadly, 150 tons is also how much plastic trash washes into the world's oceans every 9 minutes.

That's not even long enough to rock "Rock Lobster" twice, and while you shimmied with the stingray and the manta ray and the narwhal, those 300,000 pounds of polyethylene, polystyrene, polypropylene, PVC, ABS, and other polymers were rain-washed, wind-blown, or just dumped into the sea, where great gyres of plastic garbage are wreaking

an enormous dire wolf built with Salnikova from discarded plywood on-site at Maker Faire Bay Area 2018. But a life-sized blue whale? By far the most ambitious build the team had tackled. Just the weight of it, the cantilevered masses of head and body, demanded a massive steel skeleton superstructure — a first for Stockdill, but doable for local fabricators (Figure B).

But how to recycle consumer plastics to create 1,500 square feet of whale skin? That was a first for anyone. The team had never before re-manufactured waste into new building materials. The process required plenty of time, trial, and error.

"It's a strange twist on our relationship with the whales. We used to kill them for their oil — now we're turning oil-based plastics into a whale, to try and help rid the oceans of plastic." —*Joel Dean Stockdill*

havoc on marine life, entangling dolphins and whales, being ingested but not digested by turtles, birds, and fish. The problem is epic in scale, and growing.

That's why the Monterey Bay Aquarium set out to build a gigantic, 82-foot blue whale out of recycled single-use plastic and display it in front of the Golden Gate Bridge to raise public awareness. They commissioned the Building 180 arts consultancy, who brought in Joel Dean Stockdill and Yustina Salnikova to create such a thing. The team had only a vague idea of how to make a life-sized plastic whale (Figure A), but they said yes.

Monumental Commission

Stockdill's *WildLife* series of monumental animal sculptures made from refuse had graced Burning Man and festivals around the world. His new series on North American wildlife called *The Trace* includes

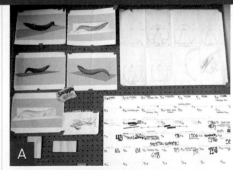

Oliver Hamilton, Keith Hammond, Shannon Riley

> "We used HDPE because it has a low melting temperature and doesn't off-gas at that temperature, making it one of the safest plastics to work with."
> —Yustina Salnikova

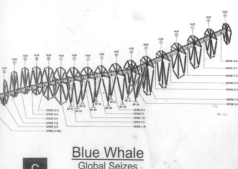

Blue Whale
Global Seizes

C

D

First they had to choose which plastic to use. Polyethylene best represented the pollution problem, because it's literally everywhere. "Polyethylene is the number one plastic used in the world, mainly HDPE #2 and LDPE #4," says Salnikova. "We used HDPE because it has a low melting temperature and doesn't off-gas at that temperature, making it one of the safest plastics to work with."

I visited Stockdill and Salnikova, then artists-in-residence at the Agapolis community high above Silicon Valley, in their cavernous open-air workshop. They walked me through the freshly assembled steel whale skeleton that towered overhead, and I asked them to show me their process for turning HDPE trash into the hundreds of massive plastic tiles that compose the whale's skin.

It's not just melt it and forget it. Here's their rough recipe:

Recycled HDPE Plastic Tiles
Makes 750+ tiles, 4–5lbs each — approx. 1,500 square feet
Enough for 1 endangered blue whale

YOU WILL NEED:
» **5,000lbs HDPE plastic** (high-density polyethylene)
» **Shears**
» **Washing machine, top loading**
» **Wood chipper, small**
» **Push stick** to jam plastic down the chipper barrel
» **Ovens** (2)
» **Baking sheet pans, large** (4)
» **Spatulas and heatproof gloves**
» **Plywood, ¾" or 1"** to build tile forms
» **2×4 lumber** to build tile press rack
» **Car jack** for tile press

1. Collect 2½ tons of HDPE plastic milk jugs, laundry detergent bottles, food barrels, etc. from your friendly neighborhood recycling center. (That's Salnikova celebrating the abundance.) Then sort out all the stray trash and non-HDPE plastics that inevitably sneak in there.

2. "Fillet" your HDPE containers into portions that will fit into the washing machine and the wood chipper.

3. Clean the fillets in the washing machine,

using the soap left over from the laundry detergent and soap bottles.

4 Shred the cleaned plastic in the wood chipper, using the push stick to cram it through. Then sort your shredded plastic by color and brightness, into enormous bins.

5 Fill 4 sheet pans with desired colors and bake in 2 ovens, 30 minutes at 350°F.

6 Don gloves and remove pans from ovens. Quickly scrape hot plastic into the plywood form and press into place by hand.

7 Close the form, slide it into the rack, then compress the form using the car jack, until cooled.

8 Remove the tile from the form, trim any extruded "flash" from the edges, and set aside.

9 Wipe sweat from eyes, and repeat 750 times. (That's Stockdill with one completed tile, about 2 square feet.)

Design to Final Build in 18 Weeks

Commissioned in April, the artists designed the whale within weeks, with structural help from Rbhu Engineering (rbhu.org) (Figure C). By June they were deep into prototyping their hand recycling process, inspired in part by the Precious Plastic series of DIY open source recycling machines (preciousplastic.com/en/machines.html). Ultimately it took 4½ months to create all the plastic panels.

Metal fabrication of the skeleton, built in 17 sections largely from 2" square steel tubing, took 4–5 weeks at Fineline Metals of Brisbane, California. The biggest joints, where the cantilevers take off, are made of massive 1⅝" steel plate I've only seen in

ships and buildings before (Figure D).

With the structure completed, it was time to weld the long curves of steel rebar to create the skin framework. The plastic skin panels were sorted by color and brightness, laid out accordingly to create light and shadow effects on the whale, and then screwed to the rebar frame using steel pipe straps and similar hardware.

During the four-day setup at San Francisco's Crissy Field, the team erected the skeleton, bolted the skin panels on, and added the eyes, mouth, and enormous chin pouch ("ventral pleats") using portions of 65 smooth, unshredded blue and white food-grade HDPE barrels (Figures E and F).

The whale, now nicknamed Ethyl and hashtagged #bigbluewhale, was unveiled Saturday, October 13, and remains on display until March 1, 2019, perhaps longer. It is shockingly huge and it looks pretty fantastic with the Golden Gate Bridge as a backdrop (Figure G).

Doing It Together

Obviously this was no ordinary DIY project, but a major DIT (do it together) campaign. From the initial impetus at Monterey Bay Aquarium, the project traveled to their ad agency, Hub Strategy and Communication. The whale idea came from Hub chief D.J. O'Neil, who had been deeply affected by the sight of a dead female blue whale on the beach in Bolinas, California, just north of San Francisco, the apparent victim of a ship strike. "You just could not believe the size of this animal," O'Neil recalls.

Hub commissioned arts consultants and producers Shannon Riley and Meredith Winner at Building 180 to find the right artists and to produce the entire build. And the ultimate installation at Crissy Field was a collaboration between the aquarium, the National Park Service, and the nonprofit Golden Gate National Parks Conservancy's Art in the Park program.

Orchestrated by Building 180, a team of 20-plus artists and uncounted volunteers worked long, dirty hours in hot summer weather at the workspace at Agapolis. Three different Bay Area recycling centers donated plastic. Rbhu Engineering, noted for its work on Burning Man mega art, made the structure safe for travel and public display. Washing all that plastic created a lot of wastewater, which the artists diverted into a Plavel Water greywater system, donated by Questa Engineering, that uses "plastic gravel" biomedia for filtration, another bit of recycling.

"Recycling Is Not Enough"

But recycling plastic won't solve the ocean pollution problem, says Kera Panni, science outreach manager for Monterey Bay Aquarium. "Recycling is not enough because plastics aren't a closed loop like glass or steel; plastic's quality degrades each time you recycle it." The demand for lower-grade plastic can't keep up with the ever-growing supply.

Some makers, led by Dutch inventor Boyan Slat, are testing ways to clean up ocean plastic (theoceancleanup.com), but even they agree the ultimate solution requires "closing the source" — preventing plastic pollution in the first place.

How can you help? Use less plastic, and repair things before you replace them. Single-use plastic is by far the main culprit. "Abandoned fishing gear is an ocean-based source that's solvable mostly by the fishing industry," Panni says. "But the majority of ocean plastics are land-based — consumer products, mostly plastic packaging. We've got to rely on reducing use."

It's hard to argue with that logic when 9 million tons of plastic go from land-to-sea each year, and the total amount in the ocean is projected to double by 2025. It's a super-sized dilemma that's crying out for more good makers to help solve it. ⊘

2019 **BOARD GUIDE**

TECH TRENDS

MICROCONTROLLERS AND SINGLE BOARD COMPUTERS GET SPECIALIZED AS THEY GROW UP

WRITTEN BY SAM BROWN

I DON'T MEAN TO MAKE YOU FEEL OLD, BUT ARDUINO COULD GET ITS LEARNER'S PERMIT TO DRIVE. For nearly 16 years now, we've had simple boards that make controlling electronics easy for anyone with the pluck to learn a few lines of code and stick some jumper wires into a breadboard. In those years, the hobby microcontroller scene has grown by leaps and bounds. Anyone can now make their own robots and inventions with just a handful of parts, a few C++ phrases, and playful persistence.

The boards of today may seem little different than the first hobby microcontrollers. Computing horsepower has bumped up a few notches. Specialized boards are offered for the maker who wants to control more things, wants a tinier board, or just one that can go in clothing without corners to snag the cloth. But at their heart, most of these boards do essentially the same thing as the originals: They check a few sensors, make a decision, and turn the power on to a few outputs in the physical world.

Yet change is in the air, and this year we're watching four trends in maker boards.

TECH TREND 1: THE PUSH FOR PYTHON

The popularity of the Python programming language grows at an amazing rate. Fans rave about how easy it is to learn, how much you can get done with very little work, and how the language makes good coding style and proper grammar the same thing: If a Python program runs, it's probably understandable.

For years, C++ has ruled the hobby microcontroller world. It makes sense: Most hobby microcontrollers have computing horsepower in the same leagues as computers from the early 1980s, and C++ is a language from 1983 that exposed every bit of power those computers could muster. Languages like Python give up some speed in exchange for all the ways they make writing programs fast and joyful. Simple computer chips like the ones at the heart of a microcontroller don't always handle a Python interpreter well. But chips get faster, year after year, including the little chips that run our gizmos.

We're seeing a raft of Python-friendly boards in this year's round-up of new microcontrollers. Leading the charge is Adafruit. In the last year they've launched Python-friendly "Express" versions of their top boards, from the tiny Trinket to the student-friendly Circuit Playground to the Arduino-like Metro and mega-sized Grand Central, and plenty more sizes in between. Whatever your project, Adafruit came out with a Python-capable board to match.

SAM BROWN
can't resist learning
and sharing what
he's learned.

Arduino Vidor 4000.

Adafruit Circuit Playground Express

Google Coral Dev Board

Nvidia Jetson AGX Xavier

CIRCUITPYTHON-CAPABLE BOARDS

Adafruit maintains an ever-growing list of boards that run CircuitPython, their branch of MicroPython (which itself works with even more boards, including micro:bit). For the full list: github.com/adafruit/awesome-circuitpython.

- **Adafruit CircuitPython boards:** Adafruit Circuit Playground Express, Adafruit Feather M0 Express, Adafruit Feather M4 Express, Adafruit Feather nRF52840 Express, Adafruit Gemma M0, Adafruit Grand Central M4 Express featuring the SAMD51, Adafruit HalloWing M0 Express, Adafruit ItsyBitsy M0, Adafruit ItsyBitsy M4, Adafruit Metro M0 Express, Adafruit Metro M4, Adafruit NeoTrellis M4, Adafruit Trinket M0

- **Arduino:** Arduino MKR 1300, Arduino MKR Zero, Arduino Zero

- **Electronic Cats:** CatWAN USB Stick, Meow Meow

- **MakerDiary:** nRF52840 Micro Dev Kit USB Dongle

- **Mini SAM:** Mini SAM development board

- **Nordic Semiconductor:** nRF52840 DK board PCA10056, nRF52840 dongle PCA10059

- **Particle:** Particle Argon, Particle Boron, Particle Xenon

- **SparkFun:** SparkFun Pro nRF52840 Mini - Bluetooth Development Board, SparkFun SAMD21 Mini Breakout, SparkFun SAMD21 Dev Breakout

TECH TREND 2:
AI EVERYWHERE

Artificial intelligence is suddenly ubiquitous. With breakthroughs in "deep learning" being applied across more and more fields, we're now seeing tasks done by computers that experts thought were still decades out of reach. And where boards that could power AI applications in a pocket gizmo were once a rarity, this year we witness a wealth of choices.

Nvidia, best known for their graphics cards, have become a darling of the AI community as the same hardware that renders graphics in stunning detail also runs and trains sophisticated neural nets. This year marked the launch of their third gadget-friendly board: The Jetson AGX Xavier. Boasting 10x the horsepower of their previous offering, the Jetson takes just 20 to 30 watts of power, a little more than one bright light bulb. That horsepower doesn't come cheap, however, carrying a price tag well north of $1,000 for the development kit.

Need to make a tiny battery pack really last, and want to do AI? Sony offered one of their AI-friendly chips up to makers this year with the launch of Spresense. Technically an MCU, Spresense doesn't quite offer the power of a single board computer, but its four cores crunch fast enough to do limited video or voice recognition, and the board runs on a shockingly low power draw of 30 milliamps, an order of magnitude less than a Raspberry Pi. Spresense demos were the first time we saw image recognition running off a pair of AA batteries.

ALSO SEE: Google's just-released Coral Dev Board is an AI-focused single board computer with hardware acceleration for TensorFlow, Google's popular deep learning software. With visual classification modeling completed in just 3ms, this is one speedy new device. (Read more on page 24.) —*Mike Senese*

TECH TREND 3:
MESH NETWORKING

Sensor meshes and swarm computing have long been a computing dream. Even before Arduino, academic labs were exploring ways to make sensor meshes work with "mote" computers, and some of that research was reborn in the Zigbee mesh networking protocol. Zigbee's now an industry standard, but most makers still haven't given it a try.

This year, Particle made a hard push into mesh networking with their new family of boards, the Argon, Boron, and Xenon (see page 34). Particle has had strong offerings for makers before: Their original board, the Photon, was one of the first consumer microcontrollers that offered built-in Wi-Fi, for a good price. What put it over the top was the friendly programming environment that automated much of the drudgery of communicating over the internet. Particle cut it down to a few lines of code, making it easy enough for beginning coders.

We found their new boards just as easy to use. The code is beginner suitable and built on the IP-based OpenThread mesh protocol. The pricing plans to manage a network are free at the scale of most personal projects, and reasonable if your invention grows to be a vast commercial success. Particle's mesh boards offer an end-to-end solution: hardware, coding, and management tools for your scattered gadgets and robots.

> **UNDER THE HOOD:** Mesh networking can be surprisingly complex, but Particle has boiled it down to a few simple rules: Every gadget in your mesh is a "leaf," a "branch," or a "gateway." Leaves and branches are both built on the inexpensive Xenon board. You make a chain of branch devices so that each branch is within Bluetooth range of the next branch in the chain, about 50 feet, tops. Each leaf needs to be in range of a branch. The branches stay powered on all the time, so they're always alert to radio messages arriving, and need a connection to a good power supply. The leaves can power down when they aren't sending in a report or requesting information, so they can stretch a small battery longer. Finally, you have a "gateway" board at the end of the chain of branches. The gateway ties the whole mesh back to the internet, using Wi-Fi if you choose the Argon board or the cellular network with the Boron board.

Will Particle's offering finally make robot swarms (see page 36) and sensor meshes easy enough for makers? Will Zigbee use take off as Digi (XBee) and Silicon Labs (Mighty Gecko) woo makers with new dev boards? Will Bluetooth Mesh eat everybody's lunch? We have our eyes on mesh networking as a technology that could dominate the maker mainstream in the next five years.

TECH TREND 4:
FPGAS — CHIPS THAT REBUILD THEMSELVES

Field Programmable Gate Arrays. FPGAs. Even the acronym blunders across the tongue. Surely someone in the marketing department could have done better, because this is a tech that, as you start to understand it, might actually blow your mind.

Computer processors are made from logic gates — little groups of transistors that make up the smallest logical operations like AND, OR, XOR, and NOT. Connecting logic gates together makes all the operations of a computer possible, much like connecting amino acids together makes all the proteins that make your body work. Logic gates are the building blocks

of the silicon chips in every microcontroller and single-board-computer you've used.

An FPGA is a chip that rearranges its own logic gates each time it powers up. It could be a microcontroller, then the next time it's powered up it could be a graphics accelerator, a bitcoin miner, or a code cracker. Each time an FPGA boots, it reads from a memory chip how its logic gates should be wired together, and that's what it becomes. Rewrite that memory store, reboot the FPGA, and it becomes a new device with a new specialty.

Several groups have tried to bring FPGAs to the maker mainstream over the last several years, but adoption has been slow. One reason is that the programming skills needed to use FPGAs are different — and more esoteric — than common programming languages like C++ or Python. Linking gates together to make a computer that does what you want takes a different kind of thinking.

Arduino waded into the FPGA fray this year with their Vidor 4000 board. The promise of the Vidor is that it should give FPGAs the same easy usage that earlier Arduino boards brought to microcontrollers. Load up an Arduino Vidor example, and it will rebuild its FPGA chip to suit, creating the hardware the example code calls for. Arduino has promised that the tools to let users make their own creations inside the Vidor's FPGA are coming soon, and will have a simple drag-and-drop interface smoothing over the classic complexity of FPGAs. (See more on page 22.)

Another FPGA board that caught our eye is the TinyFPGA BX by Luke Valenty. This is the fourth FPGA board Luke has created for the maker community. His work strikes the right balance between providing the software tools to make coding an FPGA as easy as it can be, while linking the tutorials to teach the new skills FPGAs require. While many FPGA boards are still high priced, aimed at electronics grad students and professionals, the TinyFPGA BX has a price tag similar to beginner hobby microcontrollers.

FPGAs are a cornerstone of modern electronics, replacing webs of individual chips on circuit boards with a single chip that can build its own webbing internally. Will the next five years see them become easy enough to crack into the maker mainstream? Fingers crossed, they will. ◉

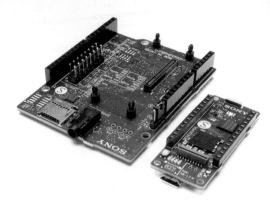

Sony Spresense

Particle Argon and Xenon

Adafruit Metro Express

Mark Madeo

TinyFPGA BX

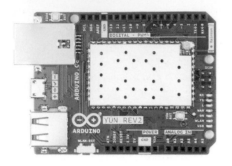

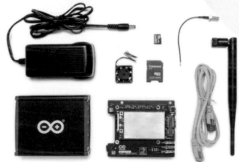

Arduino

STATE OF ARDUINO

THE ITALIAN BOARDMAKERS GET SERIOUS ABOUT FPGA AND IOT

WRITTEN BY MASSIMO BANZI

ARDUINO HAS A MISSION TO MAKE COMPLEX TECHNOLOGY SIMPLE TO UNDERSTAND AND USE. We want everyone to have the opportunity to take advantage of the possibilities it offers. From a humble beginning as a simple tool aimed at Interaction Design students, Arduino has evolved into a global community of millions of people. Initially the goal was to make it simple for people to use microcontrollers in their projects. Over time we tackled more challenging tasks like simplifying the Internet of Things, democratizing complex technologies like the FPGA, and providing better tools to educators for teaching programming, electronics, and physical computing.

THREE PILLARS

After several years of successful operations we realized that the broad audience that we serve now requires more specialization. In this respect we made several changes to our organization to reflect this need. We now look at our community through three different lenses: maker, education, and professional.

Just for the sake of clarity on the professional market, while Arduino doesn't aim to become the definitive tool for embedded engineers (they have their own tools and most likely they are capable of designing their own custom hardware), we realized that many companies already use Arduino today in order to quickly build connected devices. Their main expertise lies further down the chain — they don't want to become experts at hardware and low-level software; rather they want something that works out-of-the-box and connects securely to the cloud. They want a nice user experience which will improve their productivity. This is where they can use their expertise to create compelling applications that address real problems. We have seen this pattern many times now, with companies like Fluidintelligence

(fluidintelligence.fi) in northern Europe. They can now monitor customer machinery remotely and top-up lubricants before it becomes a problem — they are keeping machines rolling using Arduino!

IOT: NOT ONLY A HARDWARE PROBLEM

IoT is hard, as it spans a lot of areas of expertise; design, security, software, hardware, devices, gateways, cloud. But more importantly, to succeed the end application has to solve real problems — taking into account people and the environment it will operate in. Making these technologies accessible to the millions of people who could benefit from IoT was another simplification opportunity for Arduino.

A lot of work we've done in the past few years to address this challenge is coming to fruition now: We released the Arduino MKR family of open source, connected boards that support all the protocols used in professional projects — and have onboard key storage and crypto acceleration for security. We've also just released a beta of a professional-grade IoT cloud to connect these devices to — Arduino IoT Cloud. We think we met our goal to deliver a platform that would allow almost anybody to build and deploy a simple yet secure device in 5 minutes.

The list of technologies that can be made simpler is infinitely long but lately we tackled two other major milestones. We worked at making Linux more accessible to Arduino users. Today users can take any Arduino program and remotely deploy it to Linux machines including Raspberry Pi and BeagleBone with just one click. These are just some of the things now possible with Arduino Create. Thanks to the support we received from Intel you can now also remotely deploy and run Docker containers on these machines. By automating the "system administration" tasks we believe

we can make it much simpler for an Arduino user to employ and manage Linux machines in their projects.

Building on top of these capabilities we recently launched a LoRa gateway that takes advantage of an advanced radio (made by Embit) which provides enhanced LBT connectivity compared to most gateways.

We're also working on a lot of education projects aimed at helping teachers and enabling them to teach more STEAM subjects to more age groups. An example of this is the Arduino Science Kit we developed in partnership with Google which enables junior high school kids to do exciting physics experiments using the Google Science Journal app on a smartphone or a tablet.

There is still a lot of work to do with the open source community, we have a lot of tools to maintain, and we've started hiring from the community to help us manage the workload.

This past year we released the Arduino CLI, a single binary file which implements everything the classic IDE does behind the scenes. This allows people to use any development environment they like, getting the same output as the official IDE. It also enables developers to automate code compilation, uploads, and testing. You can find it on our GitHub with the other 127 repositories of open source code we share with the community. We intend to stay in the lead in open source and drive it further.

We're only at the beginning of this new journey, and we are comforted by the feedback we received from a lot of people who were waiting for a simpler (and secure) way to make their ideas a reality. We call this "Democratizing Technology," one transistor at a time. ◉

MASSIMO BANZI is the co-founder of Arduino.

MEET THE VIDOR 4000
ARDUINO'S FIRST FPGA

By Dario Pennisi, Arduino Hardware
Development Manager

The Arduino MKR Vidor 4000 hosts a number of features: onboard **8MB SDRAM, 2MB QSPI flash (1MB for user applications), micro-HDMI connector, MIPI camera connector, Wi-Fi and BLE powered by a U-Blox NINA W102 module, the classic MKR interface on which all pins are driven by both SAMD21 and FPGA, and a Mini PCI Express connector with 25 user-programmable pins.**

The FPGA features **16K of programmable logic elements, 504K of embedded RAM, and 56 18×18 bit hardware multipliers for high-speed DSP operations, such as audio and video processing. Each pin can toggle at over 150MHz and can be configured for functions such as UART, (Q)SPI, high resolution/high frequency PWM, quadrature encoders, I²C, I²S, sigma-delta DAC, etc.**

The Vidor 4000 is programmed **using the same Arduino IDE you're familiar with, both online and off.**

THE ARDUINO MKR VIDOR 4000 IS A NEW KIND OF DEVELOPMENT BOARD that combines the high performance and flexibility of an FPGA with Arduino's ease-of-use in a small MKR form factor. It contains both a Microchip SAMD21 microcontroller and a Cyclone 10 FPGA.

WHY AN ARDUINO FPGA?
The idea of introducing FPGAs in the Arduino ecosystem brings endless new possibilities, removing the barrier of fixed peripherals on CPUs and giving the freedom to use the same board for many different applications by updating the firmware.

FPGA offers the possibility to implement, in hardware, additional peripherals of any kind, from simple serial ports to complex DSP (digital signal processing) applications or even machine vision blocks.

INFRASTRUCTURE
In order to ease development, Arduino implemented an open source framework that allows users to implement functions in the FPGA and immediately have the

functionality exposed on the main processor (SAMD21). This is done by implementing a *soft processor* on the FPGA that executes the hardware driver of the implemented blocks and also provides Remote Procedure Call *APIs* through a mailbox. This provides two big benefits:

- Drivers running in FPGA offload main processor both in terms of memory and CPU load
- APIs on the main processor side are consistent and don't need to expose underlying complexity

The FPGA also contains an *auto-discovery mechanism* that allows the main processor to find out what peripherals are connected, and to which pins, issuing an error if the desired configuration is not possible (or returning the peripheral instance otherwise).

This greatly simplifies development, allowing the community to develop code that's independent from the mix-and-match of peripherals implemented in a given FPGA firmware.

To further reduce the learning curve,

Mark Madeo, Mike Senese

Arduino implemented a simple way to integrate FPGA firmware in Arduino code. A special Arduino library contains, along with headers for the instantiated peripherals, the FPGA bitstream that is downloaded with the Arduino sketch, thus preventing mismatches between code and desired FPGA firmware.

At the moment, Arduino is providing a console-based development environment with a set of open source IP blocks and some reference projects. Aside from templates, there are two main projects, Graphics and Peripherals, with the following pinout and summary information:

GRAPHICS provides the following:
- Each of the 23 MKR connector pins can be set up as a PWM pin
- 22 MKR connector pins can be set up as 11 quadrature decoder peripherals
- 11 MKR connector pins can be used to drive NeoPixel strings
- MIPI camera input (supports Raspberry PI camera v1.3)
- DVI output through micro-HDMI connector
- Compatibility with Adafruit GFX library, including several fonts, allows drawing on top of camera video and in NeoPixel frame buffers
- QR code finder block allows looking for QR codes in the image captured by the MIPI camera

PERIPHERALS provides the following:
- Each of the 23 MKR connector pins can be set up as a PWM pin
- 22 MKR connector pins can be set up as 11 quadrature decoder peripherals
- 4 MKR connector pins can be used to drive NeoPixel strings
- 8 UARTs, 4 of which can have full flow control
- 4 SPI
- 4 I²C

TEMPLATES

Along with the full-blown projects, Arduino also offers two templates, a bare-bones one that just provides a skeleton for projects that don't want to use Arduino infrastructure, and a mailbox one that provides the basic building blocks to interface FPGA blocks with the main processor, including a softcore that communicates through a mailbox.

Demo:
Object Tracking

Mini HDMI Connector

Live Text overlay (temperature)

Camera

Temperature Sensor

25.26 C

USB BLASTER

In order to support native FPGA development and debugging, Arduino implemented a sketch loadable in the main processor that emulates Intel's USB Blaster — essentially a JTAG interface that can be used along with Intel's Quartus II tool. Quartus II and USB Blaster offer functions such as FPGA programming, logic state analyzer (to check signals in the FPGA at run time) and JTAG debugging of the FPGA softcore.

FURTHER DEVELOPMENT

Arduino is developing a web-based graphical interface to allow users to create their own blend of FPGA peripherals. This was announced last year and is on track to be released by mid-2019.

Arduino is now working with the FPGA Arcade crew to port their implementation of retrogaming hardware. The group demoed an initial port of Pac-Man at Electronica 2018.

Several other FPGA IP blocks are being developed and the next IP block assembly will be focused on audio applications. ◉

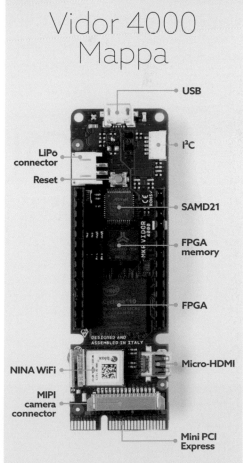

Vidor 4000 Mappa

- USB
- I²C
- LiPo connector
- Reset
- SAMD21
- FPGA memory
- FPGA
- Micro-HDMI
- NINA WiFi
- MIPI camera connector
- Mini PCI Express

DOUBLING DOWN

WRITTEN BY
BILLY RUTLEDGE

GOOGLE'S AIY ENDEAVOR LEADS TO A BRAND NEW, SELF-CONTAINED BOARD

BILLY RUTLEDGE is the director of Google's AIY and Coral teams.

GOOGLE AI IS COMMITTED TO BRINGING THE BENEFITS OF AI TO EVERYONE, and one of the ways we do this is by building tools to ensure that everyone can access AI. In 2016 I started AIY (do-it-yourself AI) as a new initiative from Google to help students and makers learn about AI through affordable, hands-on kits.

Over the past two years, we released the AIY Voice Kit and AIY Vision Kit, powered by Raspberry Pi boards and Google's AI tools to inspire the next generation of engineers. In May 2017, we released Voice Kit as a surprise extra included with *The MagPi Magazine* #57. We were humbled by the demand and the organic community growth, which resulted in the kits being picked up by Micro Center stores across the U.S. People love building their own voice interfaces, and we thought it would be great to introduce another core pillar of AI audience: image recognition. So in May 2018, we produced Vision Kit and distributed it even more broadly through Adafruit and Target stores in the U.S. As demand has grown worldwide, we've added Mouser Electronics as an international distributor.

A NEW BOARD

Since our launch, we've been thrilled with the response from students and makers. Looking at our consumers, we were excited to find industry professionals using AIY kits to prototype new product ideas with on-device AI to enhance user privacy, power efficiency, and fast performance when processing neural networks on device. So, this year we decided to invest more with this audience and offer a new platform of reliable hardware components and software tools that allow them to prototype with on-device AI, in a way that easily scales to production.

We call it Coral, a platform for experimentation for on-device AI. We chose the name because it represents an evolving, diverse ecosystem and we want Coral to foster a thriving AI ecosystem with all audiences, from makers to professionals, because we believe that great innovation happens when we all work together.

Our new products include Google's Edge TPU chip, a purpose-built ASIC that accelerates neural networks running on-device, delivering fast processing without the overhead of passing data up to the cloud. It also makes it easy to enhance privacy by keeping user data on the device, under the user's control. And, for those who need it, it works with Google Cloud's suite of Internet of Things (IoT) services to allow for remote management and easy neural network development.

Coral products currently offer the Edge TPU in two formats, as a fully integrated development board and a pluggable accessory to existing systems.

CORAL EDGE TPU DEV BOARD

Our dev board was designed for professionals who need a fully integrated system. It uses a System on Module (SoM) design where the module containing the CPU/GPU/TPU snaps into the baseboard using high density connectors. They combine to form a single-board computer for prototyping, and the SoM is available as

a stand-alone part that can be purchased in bulk for a production line.

The SoM includes the new NXP iMX8M SoC, connected to our Edge TPU over the PCIe bus. It also includes Bluetooth 5.0, dual-band 2.4/5GHz 802.11ac, and a crypto chip for secure cloud connections when needed. The default OS is called Mendel, our derivative of Debian Linux for Coral boards.

The baseboard includes a variety of connectors to make it easy to bring in sensor data and attach to peripherals, including a 40-pin GPIO header configured in a way that's consistent with many accessory boards on the market today.

THE EDGE TPU USB ACCELERATOR

Our USB Accelerator is a pluggable accessory to upgrade existing systems, for example a Raspberry Pi board. In fact, we designed the case to have the same footprint as Raspberry Pi Zero and the same mounting holes, assuming this would be a popular setup.

It has the same Edge TPU chip as the single-board computer for USB 2.0/3.0 systems. We're leading with Linux drivers first and will support other OSs soon. And a PCIe card version of the accelerator is also in the works.

We have a Python SDK to let application developers interact with the Edge TPU chip. And neural network developers will use TensorFlow Lite, which uses a smaller set of operators designed for embedded systems, along with our web hosted compiler, to produce their models.

To help you get started, we're releasing a number of sample Python applications and pre-compiled models built using open source architectures that have been tested for the Edge TPU. You'll be able to run these out of the box, re-train them for your needs, or just create your own from scratch using the software tools.

Looking ahead, we plan to expand Coral for a broad range of use cases in a way that's easy to prototype and scale to production. Our AIY kits will continue to be available, and we're looking at how to evolve them further using our Edge TPU products.

We hope you'll find our new platform inspiring for your ideas, and we're excited to see what you'll make! ⊘

1 Google's new Coral Dev Board (previous page) and Coral USB Accelerator (this page) offer stand-alone and add-on AI prototyping options.

2 A supersized, functional AIY Voice Kit with the standard version on top.

3 A tricked-out Voice Kit build.

4 The innards of the Vision Kit show the AIY bonnet atop a Raspberry Pi Zero, housed in the maker-friendly cardboard enclosure.

Mark Madeo, Google, Mike Senese

Left

Paul Trebilcox-Ruiz, Mike Rigsby, Dmitri Villevald, MeArm, Larry Lindsey, Kev Hester

AIY PROJECTS

OVER THE PAST TWO YEARS, MAKERS HAVE CREATED INNOVATIVE AND SURPRISING PROJECTS WITH THEIR GOOGLE AIY KITS. HERE ARE A FEW YOU MAY WANT TO TRY YOURSELF.

❶ TRANSLATION TOY

An animatronic toy is a great place to start hacking. In Paul Trebilcox-Ruiz's case, he used a *Star Wars* Porg as the framework for an AIY Voice-powered translation tool, designed to help youngsters learn new languages. Press the Porg's belly to put it into listening mode; it will hear your phrase, then repeat it back to you in the predetermined language while flapping its wings. hackster.io/PaulTR/translation-toy-013bfc

❷ CANDY TOSSER

This elaborate setup from Mike Rigsby has the perfect payoff — it listens for your voice (via an Amazon Echo), then checks to see if you're smiling using the AIY Vision Kit. If you are, a robot catapult points itself in the direction the Echo determined your voice to be coming from and flings some candy to you. makershare.com/projects/handy-candy

❸ HAND COMMAND RECOGNIZER

By setting up a specific search region and utilizing a training set of just 1,500 images, Dmitri Villevald constructed a program with his Vision Kit that can recognize direction-oriented hand gestures. "It can be used to control your mobile robot, replace your TV remote control, or for many other applications," he describes in his notes — and it all happens without need for the cloud. His instructions for building a companion display box, with light-up arrows that respond to your gestures, are included in the how-to. hackster.io/dvillevald/hand-command-recognizer-on-google-aiy-vision-kit-3786f6

❹ VOICE CONTROLLED ROBOT ARM

Press the AIY Voice button, speak an orientation and value (in degrees), and the affordable MeArm robot gripper leaps to do your bidding. The interface and code control all four servos on the robot for directional control. "More complex procedures, like 'pick up something in the lower left, move it to the upper right and release it there' can easily be coded and combined with a individual keyword to activate them," project creator Dr. H writes in his how-to. instructables.com/id/AIY-Voice-Controlled-MeArm

❺ YACHT TV

This project from Larry Lindsey and Christiana Caro uses the Vision Kit to identify boats that pass by their office, and save the corresponding video files for their art project, *Yacht TV* (twitter.com/YachtSee). Their documentation includes interesting notes about fine-tuning the program to increase the number of correctly identified boat sightings. It's open source and configurable — they say that just by editing a text file, you could build a Dog TV, Plane TV, you name it. hackster.io/larrylindsey/yacht-tv-366783

❻ BIRD WATCHER

Similar to *Yacht TV*, Kev Hester set up his Vision Kit to identify hummingbirds that might be visiting the feeder outside his kitchen window. When the system detects one, it snaps a photo of it and tweets it to the account twitter.com/hummingbot1. It runs on 100 lines of Python code, and he's even included the STL for the printable window bracket he designed for it. hackster.io/punkgeek/hummingbot-machine-vision-for-hummingbird-tweeting-1295f2

YACHT DETECTED

5

6

Pimoroni, Adafruit, PJRC

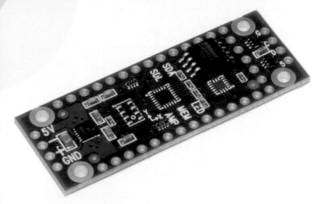

PROP UP YOUR PROJECT
ADD SOME EASY EFFECTS WITH THESE SUPER
SPECIALIZED EXPANSION BOARDS WRITTEN BY CALEB KRAFT

WHILE THE EVER-EVOLVING STATE OF DEVELOPMENT BOARDS GETS LOTS OF ATTENTION, it can be easy to gloss over their add-on boards, despite (or perhaps because of) just how many exist.

By their nature, expansion boards (called hats, capes, shields, wings, and more, depending on their platform) have a more specialized purpose than the dev boards they connect to. But there are some that are so specific that they alone nearly complete the electronics of the project for which they're meant. Here are a few special add-on boards that are perfect for those who would rather focus on the rest of the build than the circuitry inside.

PICADE X HAT FOR RASPBERRY PI
shop.pimoroni.com/products/picade-x-hat

When designing your ultimate arcade machine, you can relax and forget about the electronics. With the Picade X Hat for Raspberry Pi, all you have to do is plug in your various buttons and joysticks, leaving you to focus on designing and building your cabinet and decorations.

ANIMATED EYES BONNET
adafruit.com/product/3813

Until recently, adding digital interactive eyes to a costume would be an incredibly complex task. The Snake Eyes Bonnet from Adafruit rests on your Raspberry Pi 3 and breaks out full control of two LCD- or OLED-based eyes. Plop this into the head of your costume with a cabochon over each screen for depth and you've got a fantastic effect for virtually no effort.

PROP-MAKER FEATHERWING
adafruit.com/product/3988

Laser blasters, lightsabers, or luminous unicorn horns: Adafruit's Prop-Maker FeatherWing comes with a plethora of commonly used prop features built-in so you can save space and skip frustration. Run lights, detect motion, play sounds, all while running on battery power and without having to do any custom circuits.

TEENSY PROP SHIELD
pjrc.com/store/prop_shield.html

This is another general purpose prop tool. Plug this into your Teensy microcontroller and you've got a motion sensor, small amplifier, and LED driver ready to go. Squeeze it into your prop and you'll be up and running in no time. ✪

CALEB KRAFT is a senior editor for *Make:* magazine. His basic knowledge of electronics and code has led him to explore a multitude of options to do less circuit work, and still get the same results.

PROTOTYPING BOARDS
GET STARTED WITH THESE TRIED-AND-TRUE OPTIONS

by Jam + Walker!

microcontroller projects

LED array

Liquid dispenser

Audio synthesizer

microprocessor projects

Arcade console

Security camera

Smart home hub

ARDUINO NANO

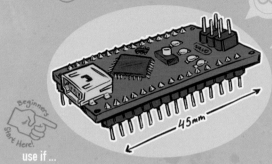
45mm

Beginners Start Here!

use if ...

- You want it to be battery powered
- You want a faster response time (vs. Pi)
- You want amazing community support! <3

RASPBERRY PI 3 MODEL A+

65mm

Linux Power!

use if ...

- You need concurrent processes
- You need a graphic interface (GUI)
- You want Ethernet or Wi-Fi

PARTICLE PHOTON

37mm

use if ...

- You're making an IoT device
- You want to control it with an app
- You will need cloud data storage

ONION OMEGA 2

Omega2
43mm

use if ...

- You want to level up your Wi-Fi game
- You need ultra-low-power Linux (0.6W!)
- You want a very small form factor

ADAFRUIT CIRCUIT PLAYGROUND

Ø50.6mm

use if ...

- You want to avoid messy wires (buttons, LEDs, buzzer, and sensors are built right into the board!)

BEAGLEBONE BLACK

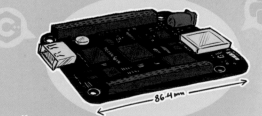
86.4mm

use if ...

- You want something open source
- You will need custom hardware
- You will want to make it in volume

by: @angelamelick and @machinehum

A **BRIEF** **HISTORY** OF **FPGA**

GET UP TO SPEED ON USING THESE ADVANCED BOARDS FOR YOUR HOBBYIST PROJECTS AND MORE

WRITTEN BY DAVID ROMANO

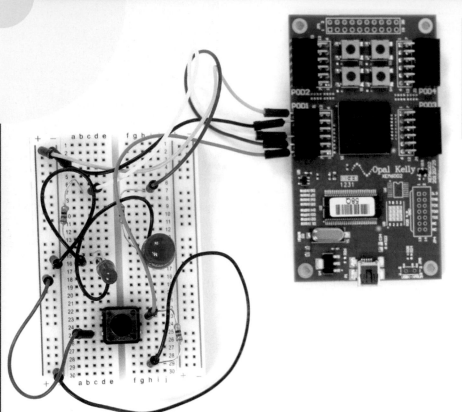

In the easy-to-follow book Make: FPGAs — Turning Software into Hardware with Eight Fun & Easy DIY Projects, *author David Romano details the ins-and-outs of the hyper-configurable field-programmable gate array (FPGA) boards and explains how — and why — to get started with these advanced devices. The following is excerpted from his book, available at* makershed.com.

INDUSTRIAL ROOTS

A *field-programmable gate array* is an integrated circuit whose fundamental hardware functionality can be programmed in the field after manufacture. There are many reasons a design team will consider FPGA technology in industry. For example, in many silicon IC design companies, FPGA-based platforms are used for what's called "shift left" testing, where a new SoC (system on a chip) device is mapped to FPGAs early in the design phase, in order to begin

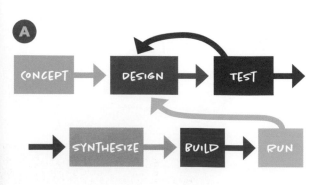

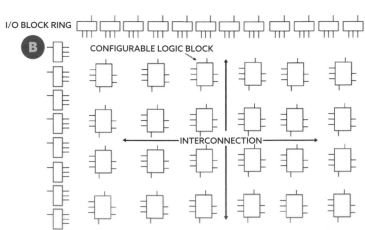

software integration long before the actual silicon device is manufactured (Figure Ⓐ). This is called "emulation" of the design. The big advantage is that emulation runs orders of magnitude faster than simulation, so you can get real-world hardware/software interactions very early in the validation phase. The FPGA system typically operates at only a fraction of the silicon operating frequency, but the time saved in integration is tremendous.

Another example of where FPGAs are considered a viable solution in industry is where the design requires having multiple hardware personalities in the same footprint. For example, this was the case for a portable test and measurement instrument that I architected when I was a design engineer. By using an FPGA in the design, the customer was able to download different test instruments to the same hardware, essentially having multiple instruments in one hardware device.

UNLIMITED OPPORTUNITIES

The real question, then, is: Why would you, the do-it-yourself hobbyist or student, even consider experimenting with FPGAs? For students, it exposes you to contemporary digital logic design methods and practices in a fun, practical, affordable way. For the hobbyist, an affordable, off-the-shelf FPGA platform can be used in some very interesting and fun DIY projects.

Many of you have had experience with Arduino or similar small microcontroller projects. With these projects, you usually breadboard up a small circuit, connect it to your Arduino, and write some code in the C programming language (which Arduino is based on) to perform the task at hand. Typically your breadboard can hold just a few discrete components and small ICs. Then you go through the pain of wiring up the circuit and connecting it to your Arduino with a rat's nest of jumper wires.

Instead, imagine having a breadboard the size of a basketball court or football field to play with and, best of all, no jumper wires. Imagine you can connect everything virtually. You don't even need to buy a separate microcontroller board; you can just drop different processors into your design as you choose. Now that's what I'm talking about! Welcome to the world of FPGAs.

FPGA HISTORY

Xilinx Inc. was founded in 1984, and as the result of numerous patents and technology breakthroughs, the company produced the first family of general-purpose, user-programmable logic devices based on an array architecture. It called this technology breakthrough the *Logic Cell Array (LCA)*, and with this the Xilinx XC 2000 family of FPGAs was born.

You can think of an LCA as being made up of three types of configurable elements: *input/output (I/O) blocks*, and *logic blocks*, and an *interconnect matrix* (Figure Ⓑ). From these, a designer can define individual I/O blocks that interface to external circuitry. You can think of these as configurable pins of ports. The designer can also use logic blocks, connected together through the interconnect matrix, to implement logic functions. These functions can be as simple as a counter or as complex as a microcontroller core. In a way, the interconnect matrix is like the wires on a breadboard that connect everything together — but completely programmable.

Before there were FPGAs, you needed to use dozens of discrete ICs on a circuit board, or sometimes even hundreds of ICs on multiple circuit boards, to accomplish the hardware functionality you can achieve today with one FPGA device.

For example, today you can create the entire Pac-Man arcade game on a single FPGA device, including the game software. Now that's fun!

The configuration of an FPGA device is accomplished through programming the *memory cells*, which determine the logic functions and interconnections. In the early days, this program (or what has become known as the *bit file*) was loaded at power-up from EEPROM, EPROM, or ROM on the circuit board, or loaded from a PC through a serial connection on the board from the FPGA programming tool. Since the underlying technology is volatile static RAM (SRAM), the bit file must be reloaded with every power cycle of the device.

In today's FPGAs, SD flash memory replaces the EPROM, and USB or JTAG replaces the serial connection, but the programming function remains much the same as it was in the beginning. ⊘

David Romano

DAVID ROMANO is the president and CEO of the STEM education company Tri-Tech Pathways Inc. He is pursuing a doctorate degree in education.

MINI MAC

TIME REQUIRED:
2–3 Hours

DIFFICULTY:
Easy

COST:
$40

MATERIALS
» **HalloWing M0 Express microcontroller board** Adafruit #3900 adafruit.com
» **Speaker, mini oval, 8Ω, 1W** Adafruit #3923
» **Battery, lithium polymer (LiPo), 500mAh (optional)** Adafruit #1578
» **3D printed enclosure parts**

TOOLS
» **3D printer, FDM type**
» **STL files, digital assets, and CircuitPython code** available on the project page, learn.adafruit.com/hallowing-mac

PRINT A MINUTE MACINTOSH SE AND STUFF A HALLOWING INSIDE FOR RETRO NOSTALGIA WRITTEN AND PHOTOGRAPHED BY NOE AND PEDRO RUIZ

THIS TINY MACINTOSH SE DISPLAYS A MAC BOOT SCREEN AND PLAYS THE CLASSIC STARTUP CHIME SOUND. It uses the Adafruit HalloWing M0 Express microcontroller board and runs CircuitPython, which makes it easy to display images on the HalloWing's screen. The capacitive touch pads cycle through images and sounds stored on the device.

1. POWER UP
For our build, USB power is used. Optionally, a 500mAh LiPo battery can be used to make it portable — it fits nicely inside the case and it's rechargeable via the USB port on the HalloWing.

2. LOAD THE CODE
The code for this project was developed using Adafruit's CircuitPython and documented in John Park's Tiny Museum project (learn.adafruit.com/tiny-museum-tour-device/code-with-circuitpython). Follow his tutorial to install and set up CircuitPython on your HalloWing.

3. UPLOAD MAC ASSETS
Once you have CircuitPython set up, you can upload the assets for this project. These include bitmap images of the Mac OS boot screens and chime audio WAV files. With the board mounted on your computer, simply drag the *.bmp* and *.wav* files onto your HalloWing's *CIRCUITPY* drive. These must be at the top level of your HalloWing, not inside a folder. Also make sure the *code.py* and *adafruit_slideshow.mpy* files have been copied over.

4. PRINT THE MAC ENCLOSURE
The printable STL files are oriented to print "as is" — including the *hallowmac-case-back* that features a back door panel with a print-in-place hinge (Figure 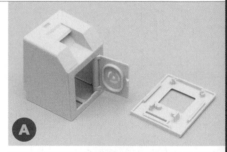). This allows the panel to swing open and closed. These parts require tight tolerances; you might need to adjust your slicer settings.

5. ASSEMBLY
Snap the HalloWing onto the faceplate (Figure ). It mounts without screws.

A mini speaker plugs into the audio port on the back of the HalloWing. Press-fit the speaker into the back door and close it with the built-in latch.

6. FIRE IT UP!
Reach inside to trigger the capacitive touch pads for advancing to the next or previous image. The code looks for *.bmp* files to display — when it finds one it then parses that filename and looks for a *.wav* file with the same name (Figure ). For example, **macos8.bmp** causes the code to look for a **macos8.wav** file to play. Enjoy!

[+] Find more tips and instructions on the project page: learn.adafruit.com/hallowing-mac.

NOE AND PEDRO RUIZ are part of the Adafruit team, making projects and hosting a weekly live stream show called 3D Hangouts.

Make:

THE *ORIGINAL*
GUIDE TO
BOARDS

2019
accept no imitations!

Brought to you by

BOARDS BOARDS BOARDS!

Welcome to the 2019 Guide to Boards! It's often said, when you're looking to make a project, be sure to use the right tool for the job — this applies as much to woodworking as it does to a high-powered electronics endeavor. With that in mind, this guide is designed to help you find the perfect brain for your creation. We've gathered and listed the specs of nearly all the latest and greatest boards available now, including microcontrollers, single board computers, and FPGAs. From robotics to AI to IoT, you'll find what you need here. Dig in! —*Mike Senese*

MICROCONTROLLERS (MCU)

Board Name	Price	Dimensions	Software	Clock Speed	Processor	Memory	Digital Pins	Analog
Adafruit Circuit Playground Express	$25	2" dia.	CircuitPython, MakeCode, Arduino IDE, code.org CS Discoveries	48MHz	32-bit ATSAMD21G18 Cortex-M0	256KB flash, 32KB RAM, 2MB SPI flash	8	8
Adafruit Feather 32u4 Bluefruit LE	$30	2"×0.9"	Arduino, C	8MHz	8-bit ATMega32U4	32KB flash, 2KB RAM	20	6
Adafruit Feather 32u4 RFM96 LoRa Radio - 433MHz	$35	2"×0.9"	Arduino	8MHz	8-bit ATMega32U4	32KB flash, 2KB RAM	20	10
Adafruit Feather Huzzah	$17	2"×0.9"	Arduino, C, MicroPython, JavaScript. Lua	80MHz	32-bit ESP8266	4MB flash	9	
Adafruit Feather M4 Express	$23	2"×0.9"	Arduino, C, Circuit-Python, MakeCode	120MHz	32-bit ATSAMD51 Cortex M4	512KB flash, 192KB RAM, 2MB SPI flash	15	
Adafruit Flora	$15	1.8" dia.	Arduino	8MHz	8-bit ATmega32U4	32KB flash	1–10	4-
Adafruit Gemma M0	$10	1.1" dia.	Arduino, C, Circuit-Python, MakeCode	48MHz	32-bit ATSAMD21	256KB	3	3 (1
Adafruit Grand Central M4 Express	$38	4"×2.1"	Arduino, C, Circuit-Python, MakeCode	120MHz	32-bit ATSAMD51P20 Cortex-M4	1MB flash, 256KB RAM, 8MB QSPI flash	46	
Adafruit Huzzah32	$20	2"×0.9"	Arduino	240MHz	Dual core Tensilica LX6	4MB flash, 520KB SRAM	8–11	6
Adafruit Metro M0 Express	$25	2.8"×2"	Arduino, C, Circuit-Python, MakeCode	48MHz	32-bit AT-SAMD21G18 Cortex-M0	256KB	24	6 (.
Adafruit Metro M4 Express	$28	2.8"×2.1"	Arduino, C, Circuit-Python, MakeCode	120MHz	32-bit ATSAMD51J19 Cortex-M4	512KB flash, 192KB RAM, 2MB QSPI flash	17	
Adafruit NeoTrellis	$40	4.7"×2.4"	Arduino, CircuitPython	120MHz	32-bit ATSAMD51G Cortex-M4	512KB flash, 192KB RAM, 8MB QSPI flash	0	
Adafruit Trinket 3.3V & 5V	$7	1.1"×0.6"	Arduino	8MHz or 16MHz	8-bit ATtiny85	8KB flash	5	
Arduino Mega	$39	4"×2.1"	Arduino	16MHz	8-bit ATMega2560	256KB flash	50+	
Arduino MKR WAN 1300	$40	2.7"×1"	Arduino	48MHz	32-bit SAMD21 Cortex-M0+	256KB flash, 32KB RAM	8–15	
Arduino MKR WiFi 1010	$34	2.5"×1"	Arduino	48MHz	32-bit SAMD21 Cortex-M0+	256KB flash, 32KB RAM	8	
Arduino Uno	$22	2.7"×2.1"	Arduino	16MHz	8-bit ATMega-328PU	32KB flash	11–20	
Bare Conductive Touch Board	$80	3.3"×2.4"	Arduino	16MHz	8-bit ATMega32U4	32KB flash, microSD	11–20	
BBC micro:bit	$15	1.97"×1.57"	MakeCode, Python	16MHz	32-bit ARM Cortex-M0	16KB RAM	11–20	
Bolt WiFi Module	$22	1.4"×1.4"	JavaScript	80MHz	32-bit Tensilica Xtensa LX106	64KB flash, 96KB RAM	5	
DFRobot Leonardo with Xbee	$20	2.8"×2.2"	Arduino	16MHz	8-bit ATmega32U4	32KB flash	11–20	7

For more boards and reviews, visit our online Makers' Guide to Boards: makezine.com/comparison/boards

... and watch the @make and @digikey channels for info on accessing upcoming augmented features of this guide.

Radio	Video	Ethernet On Board	Input Voltage	Battery Connection	Operating Voltage
—	—	—	5V	✓	3.3V
Bluetooth	—	—	3.7V–5V	✓	3.3V
LoRa	—	—	5V	✓	3.3V
Bluetooth	—	—	3.7V–5V	✓	3.3V
—	—	—	3.3V	✓	3.3V
—	—	—	3.5V–16V	✓	3.3V
—	—	—	3V–6V	✓	3.3V
—	—	—	6V	—	3.3V
Bluetooth	—	—	3.3V–5V	✓	3.3V
—	—	—	6V–12V	—	3.3V
—	—	—	6V	—	3.3V
—	—	—	5V	—	3.3V
—	—	—	3.3V–16V	—	3.3V or 5V
—	—	—	6V–20V	—	5V
Ra	—	—	5V	✓	3.3V
i-Fi	—	—	3.7V–5V	✓	3.3V
—	—	—	6V–20V	—	5V
—	—	—	3V–5.5V	✓	5V
tooth	—	—	1.8V–3.3V	✓	3.3V
Wi-Fi	—	—	5V	—	3.3V
Bluetooth	—	—	6V–12V	—	5V

Mark Madeo

NEW AND NOTABLE
By Sam Brown

ADAFRUIT FEATHER M4 EXPRESS

The Feather M4 Express is at the crossroads of Adafruit's two recent pushes: It has the Feather form factor that makes it breadboard-friendly with plenty of plug-in boards to add to its features. And it has the ATSAMD51 processor, giving you the option to code it in classic C++, or Python. A spot of perf board at the end of the Feather even gives you room to solder on a few extra components, to hold whatever last few bits your project needs.

ARDUINO MKR WIFI 1010

Does the Uno finally have competition as Arduino's most beginner-friendly board? Arduino's new MKR format is more breadboard-friendly than the original, with stacking headers so you have the option to seat it in a breadboard or wire straight in. The MKR WIFI 1010 adds Wi-Fi, so the projects you build on it can be internet enabled.

RASPBERRY PI 3 MODEL A+

Raspberry Pi returns to its roots with the model A+, now in the middle of RPi's offerings between the I/O-rich model B+ and the tiny Pi Zero line. This one's a just-right fit for embedded computers that need a bit more horsepower than the Zero can muster, but don't need all the desktop-friendly features of the B+, like the added USB jacks and memory.

MICROCONTROLLERS (MCU)

Board Name	Price	Dimensions	Software	Clock Speed	Processor	Memory	Digital Pins	Analog Pins
Digilent uC32	$35	2.7"×2.1"	Arduino, MPIDE	80MHz	32-bit PIC32MX340F512H	512KB flash, 32KB RAM	42	12
Digilent Wi-Fire	$79	3.5"×2.1"	Arduino, MPIDE	200MHz	32-bit PIC32MZ2048E-FG100	2MB flash, 512KB RAM	43	12
Espruino Pico	$25	1.3"×0.6"	Espruino JavaScript Interpreter	84MHz	32-bit Cortex-M4	384KB flash, 96KB RAM	22	9
Espruino WiFi	$35	1.2"×0.9"	Espruino JavaScript Interpreter	100MHz	32-bit Cortex-M4	512KB flash, 128KB RAM	21	8
LilyPad Arduino USB - ATmega32U4	$26	2" dia.	Arduino	16MHz	8-bit ATMega32U4	32KB flash	5	4
Netduino 3 WiFi	$50	2"×3"	C# .Net	168MHz	32-bit Cortex-M4	1408KB flash, 164KB RAM	22	6
Nordic Thingy:52	$40	2.4"×2.4"	Nordic Thingy	64MHz	nRF52832	512KB flash, 64KB RAM	12–30	12–
Particle Argon	$25	2"×0.8"	Particle Device OS	64MHz w/ a 240MHz coprocessor	32-bit Cortex-M4F w/ ESP32 as Wi-Fi coprocessor	1MB flash, 256KB RAM	7	6
Particle Electron	$49 for 2G, $69 for 3G	2"×0.8"	Arduino	100MHz	32-bit STM32F205 Cortex-M2	1MB flash	21–50	7–1
Particle Photon	$19	1.44"×0.8"	Arduino	100MHz	32-bit STM32F205 Cortex-M3	1MB flash	11–20	7–1
Particle Xenon	$15	2"×0.8"	Particle Device OS	64MHz	32-bit Cortex-M4F	1MB flash, 256KB RAM	7	
PJRC Teensy 3.2	$20	1.4"×0.7"	Arduino	32MHz	32-bit MK20DX256 Cortex-M4	256KB flash	21–50	
PJRC Teensy 3.5	$25	2.5"×0.7"	Arduino	120MHz	32-bit Cortex-M4	1MB flash, 256KB RAM, 4K EEPROM	37–62	2
PJRC Teensy 3.6	$29	2.5"×0.7"	Arduino	180MHz	32-bit Cortex-M4	1MB flash, 256KB RAM, 4KB EEPROM	37–62	2
Realtek RTL8710	$4	0.945"×0.63"	FreeRTOS	32MHz	32-bit Cortex-M3	1MB flash, 512KB RAM	11–20	
Seeed LinkIt 7697	$15	1.9"×1"	LinkIt SDK using GCC, IAR embedded workbench, KEIL uVision	192MHz	32-bit Cortex-M4	4MB flash, 352KB RAM	18	
Sino:bit	$29	3.615" (octagon)	Other	16MHz	32-bit Cortex-M0	256KB flash, 16KB RAM	8–11	
SparkFun Blynk Board	$32	2"×1.66"	Blynk App	16MHz	32-bit ESP8266EX	4MB flash, ~50KB RAM	1–10	4
SparkFun Digital Sandbox	$50	4.175"×2.95"	Arduino	8MHz	8-bit ATMega328	32KB flash, 2KB RAM	1–3	
SparkFun ESP32 Thing	$22	2.35"×1"	ESP-IDF toolchain, Arduino	240MHz	32-bit ESP32	16MB flash, 520KB RAM	28	1
SparkFun RedBoard Edge	$22	2.7"×2.1"	Arduino	16MHz	8-bit ATMega328PU	32KB flash	14 (4 PWM)	
SparkFun RedStick	$21	2.6"×0.7"	Arduino	16MHz	8-bit ATMega328P	32KB flash	14 (6 PWM)	
Spresense	$65	1.96"×0.81"	NuttX emulating Arduino	156MHz	32-bit Cortex-M4F x 6 cores	1.5MB SRAM, 8MB flash	17, extension 14	2 AD slo
Stemtera Breadboard	$45	4.4"×3.1"	Arduino	16MHz	8-bit ATMega328P and 8-bit ATMega32U2	32KB flash, 2KB RAM	11–20	
Texas Instruments TM4C1294XL	$21	4.9"×2.2"	Energia, Code Composer, others	100MHz	120MHz 32-bit Cortex-M4	1MB flash, 256KB RAM, 6KB EEPROM	21–50	
Thunderboard Sense	$36	1.77"×1.18"	Simplicity Studio	32MHz	32-bit ARM Cortex-M4 core with 32KB RAM	256KB flash	11–20	
TinyLily Mini	$10	0.55" dia.	Arduino	8MHz	8-bit ATMega328P	32KB flash	1–10	
WiPy	$30	1.7"×1"	Arduino	32MHz	32-bit TI CC3200 Cortex-M4	2MB flash	21–50	1–

Radio	Video	Ethernet On Board	Input Voltage	Battery Connection	Operating Voltage
–	–	–	7V–15V	–	3.3V
Wi-Fi	–	–	7V–15V	–	3.3V
–	Composite and VGA	–	3.3V–16V	–	3.3V
Wi-Fi	Composite and VGA	–	3.7V–5V	–	3.3V
–	–	–	2.7V–5.5V	✓	3.3V
Wi-Fi	–	–	5V–12V	–	3.3V
Bluetooth	–	–	5V	✓	3.3V
Wi-Fi	–	–	4.5V–5.5V	✓	3.3V
–	–	–	3.9V–12V	✓	3.3V
Wi-Fi	–	–	3.6V–5.5V	–	3.3V
...etooth	–	–	4.5V–5.5V	✓	3.3V
–	–	–	3.6V–6V	–	3.3V
–	–	–	5V	–	5V
–	–	–	5V	–	3.3V
...i-Fi	–	–	5V	–	3.3V
...Bluetooth	–	–	5V	–	3.3V
...uetooth	–	–	1.8V–3.3V	✓	3.3V
...Wi-Fi	–	–	3.7V–6V	✓	3.3V
–	–	–	5V	✓	5V
..., Bluetooth	–	–	5V	✓	5V
...Bluetooth	–	–	6V–20V	–	5V
–	–	–	2V–6V	–	5V
–	–	–	5V	–	Main Board: 1.8V, 0.7V ADC inputs, powered by a 3.7V battery
–	–	–	6V–20V	–	5V
–	–	✓	5V	–	3.3V
...uetooth	–	–	2V–5.5V	–	3V
–	–	–	2.7V–5.5V	–	3V
...Wi-Fi	–	–	3.3V–5.5V	–	3.3V

Mark Madeo

NEW AND NOTABLE
(continued)

ASUS TINKER BOARD S

A single board computer from a storied maker of overclockable desktop computers, the Tinker Board does not skimp on horsepower. This year sees the release of the new model "S," with hardware tweaks, like 16GB of non-volatile RAM, so you can load your OS and software package on the board and reclaim your SD card. The classic Tinker Board got a price drop to differentiate it from the model "S," giving us another reason to love this board.

POCKETBEAGLE

It's inexpensive. It fits in a small mint tin. It has 52 (52!) GPIO pins to interact with the physical world. The main processor runs at 1GHz, with a half-gig of memory for running whatever Linux programs you have in mind. It has two "PRU" subprocessors that react in a single clock cycle; great for tending sensors and keeping PWM motors running smoothly even while the main processor is loaded with the heaviest tasks you can heap on it.

TINYFPGA BX

This is a fantastic board for learning to program FPGA chips that rewire themselves. Small and inexpensive, the TinyFPGA BX has power enough for maker-scale projects. The toolchain for programming this board is 100% open source, so you don't have to wait for your license to be approved to start creating. The tutorials start you at step one, neither hiding nor exaggerating the complexities of hardware description languages.

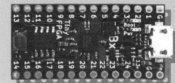

SINGLE-BOARD COMPUTERS (SBC)

Board Name	Price	Dimensions	Software	Clock Speed	Processor	Memory	Digital Pins	Analog Pins
Asus Tinker Board	$58	3.37"×2.125"	Debian Linux	1.8GHz	64-bit RK3288	2GB Dual Channel DDR3	40	—
Asus Tinker Board S	$85	3.37"×2.125"	Debian Linux (Linaro), Android 6 & 7	1.8GHz	32-bit Rockchip RK3288	2GB Dual Channel	26	3 PWM
Banana Pi M2 Berry	$36	3.6"×2.4"	Linux	1GHz	32-bit quad-core Cortex-A7 V40	1GB DDR3 SDRAM	26	—
BeagleBoard PocketBeagle	$25	2.2"×1.4"	Linux	1GHz	32-bit Cortex-A8	512MB DDR3	40	8ADC (6 1.8V, 2 3.3V)
BeagleBoard-X15	$270	4"×4.2"	Linux	1GHz	32-bit AM5728 Cortex-A15	4GB 8-bit eMMC	50+	—
BeagleBone Black	$55	3.4"×2.1"	Linux	1GHz	32-bit AM335X Cortex-A8	4GB eMMC	50+	7—
BeagleBone Blue	$82	3.4"×2.1"	Debian Linux with Cloud9 IDE and libroboticscape	1GHz	32-bit Cortex-A8, Cortex-M3, TI Programmable Real-Time Units	512MB RAM, 4GB eMMC flash	8	4
Google Coral Edge TPU Dev Board	$150	3.3"×2.2"	Mendel Linux, Android	1.3GHz	64-bit i.MX8MQ	1GB	28	—
Nvidia Jetson AGX Xavier	$1,299	4.2"×4.2"	Linux-based JetPack SDK	2.26GHz	64-bit 8 Core ARM v8.2 CPU	16GB 256-bit LPDDR4x	23	3 PW
Nvidia Jetson TX2 Dev Kit	$570	6.7"×6.7"	Linux-based JetPack SDK	2GHz	64-bit ARMv8	8GB L128-bit DDR4	6	
ODROID-XU4	$59	3.3"×2.3"	Linux	2GHz	32-bit Samsung Exynos5422/32-bit octa-core Cortex-A15	2GB	25	
Onion Omega2+	$15	1.7"×1"	Linux	580MHz	32-bit MIPS	128MB	18	
Qualcomm DragonBoard 410c	$75	2.12"×3.35"	Android, Linux, Win 10 IoT	1.2GHz	64-bit Snapdragon 410	1GB LPDDR3 533MHz, 8GB flash	12	
Raspberry Pi 3, Model B+	$35	3.4"×2.2"	Linux	1.4GHz	64-bit Broadcom BCM2837B0, quad-core A53 (ARMv8)	1GB LPDDR2	29	
Raspberry Pi 3, Model A+	$25	2.6"×2.2"	Raspbian Linux	1.4GHz	64-bit Broadcom BCM2837B0, quad-core A53 (ARMv8)	512MB	24	
Raspberry Pi Zero W	$10 (non-wireless version: $5)	1.18"×2.56"	Linux	1GHz	32-bit Broadcom ARMv6	microSD	21–50	
UDOO	$65	3.5"×2.3"	Linux	1GHz	32-bit Freescale i.MX 6SoloX ARM Cortex-A9	microSD	24	
VoCore 2	$18	1"×1"	Linux	100MHz	16-bit MT7628AN, 580MHz, MIPS 24K	128MB, DDR2, 166MHz	21–50	4

FIELD-PROGRAMMABLE GATE ARRAY BOARDS (FPGA)

Board Name	Price	Dimensions	Software	Clock Speed	Processor	Memory	Digital Pins	Analog
Alchitry Au	$110	2.56"×1.77"	Alchitry Labs, Vivado	100MHz oscillator on board	32-bit XC7A35T-1FTG256C	256GB DDR3, 32MB flash	102	
Alchitry Cu	$55	2.56"×1.77"	Alchitry Labs, Vivado	100MHz oscillator on board	iCE40-HX8K	32MB configuration flash	79	
Arduino MKR Vidor 4000	$60	3.27"×1"	Arduino	48MHz and up to 200MHz	32-bit ATSAMD21 Cortex-M0+	256KB flash, 32KB SRAM, (FPGA 2MB flash and 8MB SDRAM)	22 and 25 pin Mini PCI Express header	
TinyFPGA AX2	$18	1.2"×0.7"	Verilog	133MHz	XO2-1200	8KB flash, 9KB RAM	21	
TinyFPGA BX	$38	1.4"×0.7"	Verilog	16MHz	iCE40-LP8K	8 MBit flash, 128KB RAM	41	

Radio	Video	Ethernet On Board	Input Voltage	Battery Connection	Operating Voltage
Wi-Fi, Bluetooth	HDMI	✓	5V	–	5V
Wi-Fi, Bluetooth	HDMI	✓	5V	–	5V
-Fi, Bluetooth	HDMI	✓	5V	–	3.3V
–	–		5V	–	3.3V
–	HDMI	✓	12V	–	3.3V
–	Micro-HDMI	✓	5V	–	1.8V and 3.3V
i, Bluetooth	–	–	9V–18V	✓	1.8V–7.4V
-Fi, Bluetooth	HDMI	✓	5V	–	5V
–	HDMI, eDP 1.2a, DP 1.4	✓	9V–20V	–	1.8V
i, Bluetooth	HDMI	✓	6.75V–19V	–	19V
–	HDMI	✓	5V	–	1.8V
Wi-Fi	–	–	5V	–	3.3V
Bluetooth	HDMI	–	6.5V–18V	–	1.8V
Bluetooth	HDMI	✓	5V	–	3.3V
Bluetooth	HDMI	–	5V	–	3.3V
Bluetooth	Micro-HDMI	–	5V	–	3.3V
Bluetooth	Micro-HDMI	✓	6V–15V	–	3.3V
Wi-Fi	–	✓	3.6V–6V	–	3.3V

Radio	Video	Ethernet On Board	Input Voltage	Battery Connection	Operating Voltage
–	–	–	5V	–	3.3V and 1.8V IO, 1V FPGA core, 1.35V DDR3L, 1.8V analog
–	–	–	5V	–	3.3V IO, 1.2V FPGA core
...uetooth	Micro-HDMI	–	5V	✓	3.3V
–	–	–	3.3V	–	3.3V
–	–	–	5V	–	3.3V

GENESIS OF A BOARD

By Limor Fried, founder and CEO, Adafruit

Hep Svadja

Designing and producing microcontrollers isn't just my job, it's my passion. I've been creating electronics for 18 years, starting with my very first kit, the MiniPOV, up to the newest board, Adafruit Grand Central Express. Every engineer and electronics maker has an individual process of coming up with a new product; here's what I do for mine.

1. Most boards start by watching new chips from companies, and listening to what customers and fans are looking for.

2. I study component specs and samples to determine value for beginners and experts alike.

3. From there, we design the board layout — I use Eagle CAD, but some folks here like KiCad.

4. Next, we hand-assemble a PCB prototype. Usually there are 2–4 revisions depending on complexity.

5. Tester design! We make sure to test every element of the board.

6. Production: We do an initial run of about 100 boards to start. Once those sell, we do 250 pieces, and then finally 1,000 at a time.

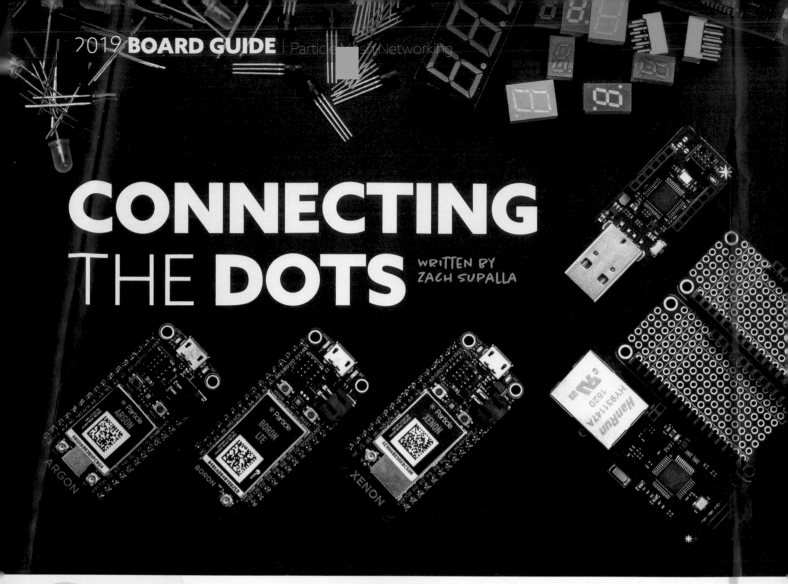

CONNECTING THE **DOTS**

WRITTEN BY ZACH SUPALLA

HOW — AND WHY — PARTICLE DEVELOPED A MESH-NETWORKABLE LINE OF MICROCONTROLLERS

ON JANUARY 29, 2012, I WAS IN MY APARTMENT IN EVANSTON, ILLINOIS, surrounded by jumper cables and electronics components. I was in graduate school, putting together a prototype of my "big idea" — an internet-connected light bulb. To prove out the basic concept, I connected an LED to an Arduino, and I connected the Arduino to the web (logged here: makezine.com/go/spark-prototype). It worked. At that moment, I became a maker. And the experience I had building that prototype sent me down an entrepreneurial journey that played out over the following seven years. Today, the company that began as that blinking LED has built the most widely used and fastest-growing platform for IoT devices in the industry.

Our goal at Particle is simple: We help people and companies connect physical devices to the web. Building an "Internet of Things" (aka IoT) product is hard, and we strive to make it easy. By doing so, we've been able to help 160,000 engineers, developers, students, artists, and designers build prototypes of IoT products, and we've helped hundreds of companies — from Fortune 500 enterprises to tiny startups — deploy IoT products at scale.

We've been successful because we embrace and tackle for our customers the hard problems that create barriers to building an IoT product. And there are a lot of them. Hardware is hard. Radios are hard. Networking is hard. Security is hard. Our goal is to gobble up those hard problems so that the product creator can focus on their product, and not the many layers of infrastructure underneath.

In 2015, this approach led to the launch of the Electron, our first foray into cellular connectivity. Building cellular-connected IoT products is particularly hard because you have to deal with cellular carriers, and they do not want to deal with you. If you're smaller than Samsung, carriers would rather not put up with you. And basically everybody is smaller than Samsung. So we launched the Electron to make it easier for everyday people and normal-sized companies to build cellular-connected IoT products.

Once we had the Electron and the Photon (our pre-existing Wi-Fi product), we found that we could help people connect IoT products wherever there was an existing wireless network.

But what about where there isn't a wireless network? What about basements? What about mines? What about forests

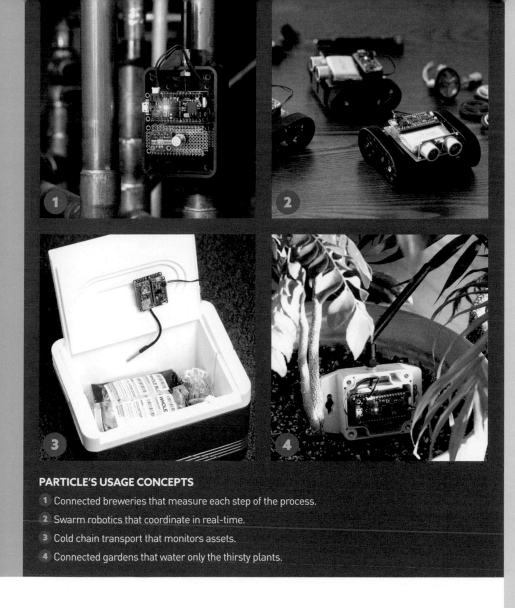

PARTICLE'S USAGE CONCEPTS

1 Connected breweries that measure each step of the process.

2 Swarm robotics that coordinate in real-time.

3 Cold chain transport that monitors assets.

4 Connected gardens that water only the thirsty plants.

and farms, far from the city and its many cellular towers?

This is why we built Particle Mesh, our solution to help product creators build their own wireless networks. Each of our new developer kits — the Argon, Boron, and Xenon — can create, manage, and deploy reliable and secure local Mesh networks that self-heal. In doing so, they help us plug the holes in our wireless world, and help you connect IoT projects and products where you couldn't before.

If you've ever tried to tweak the settings of your Wi-Fi router so that you can get a signal in the farthest corners of your house, you know that wireless networking is hard. It took ¾ of our engineering team almost a year and dozens of very patient and helpful early customers to make Particle Mesh "just work." But, in the end, we're proud to have

built something that anyone can use to build a wireless mesh network for IoT products — no Ph.D. required.

Some of the biggest problems that we can solve with IoT happen in the hidden spaces that we don't think about. Flooding starts in the dark, damp basement you'd like to forget is there; equipment failure begins in the oily underbelly of a machine shop where that old generator lives; dirty water comes from the pipes and pumps that are only accessible after a call to the water department and an excavation crew. Real solutions happen in these unlikely and oft unvisited spaces and places. Particle Mesh takes us there — so that you can begin your own IoT journey by blinking an LED. ◉

ZACH SUPALLA is the founder and CEO of Particle (particle.in)

WRITTEN AND
PHOTOGRAPHED BY
BRANDON SATROM

DIY SWARMBOTS!

TIME REQUIRED:
A Weekend

DIFFICULTY:
Intermediate

COST:
$130–$150

MATERIALS
» **Thunder Tumbler R/C cars (2 or 3)** as low as $10 each at many drug, discount, and department stores
» **Particle Argon microcontroller** Wi-Fi and mesh network gateway, store.particle.io/products/argon-kit
» **Particle Xenon microcontrollers (3)** Bluetooth and mesh network endpoint/repeater, store.particle.io/products/xenon-kit
» **Breadboards (3)** included with the Argon and Xenon kits
» **LiPo batteries (3)** 3.7V 1200mAh, such as Adafruit #258, adafruit.com
» **Assorted jumper wires** for connecting each Xenon to the R/C car PCB

TOOLS
» **Multimeter** for inspecting IC pin functions, as these may vary from one cheap R/C car to the next
» **Soldering iron and solder**
» **Screwdriver**
» **Computer with internet access**

BUILD A MESH-NETWORKED MOB OF R/C CARS WITH PARTICLE

HACKING A TOY IS A FUN WAY TO SEE HOW REAL-WORLD ELECTRONICS ARE DESIGNED AND BUILT, and to tinker freely without ruining expensive gear. And cheap consumer electronics generally are a fantastic platform for learning and even building new, innovative solutions. So when *Make:* asked me for a DIY swarmbots project, I knew just where to go.

Radio-controlled (R/C) vehicles are a common target for makers, and cars with the Thunder Tumbler name provide an affordable, accessible platform for R/C car hacking. The Tumbler has been hacked a number of times before, but I don't think it's ever been used in a mesh network. That is, until now. Using Particle's new, mesh-ready hardware, I created a swarmbot network of Tumblers that move in synchronized fashion at my every command!

A LESSON ABOUT MESHIN'

How does *mesh networking* work? Most connected solutions rely on Wi-Fi or cellular networks for connectivity. This typically means that each device maintains its own connection to the internet. While this is useful for accessing the cloud for data storage or processing, sometimes you just want your gizmos to connect with other devices locally, regardless of whether an internet connection exists.

Mesh networking enables these scenarios by allowing you to create local networks of connected devices. The bulk of the network consists of *endpoints* that sense or actuate, and *repeaters* that increase the size and reliability of the mesh by passing messages between devices. In addition, a small number of devices — often just one — serve as *gateways* to maintain a connection

to the internet. Critically, these local networks of devices can still communicate with each other when the internet connection disappears. For jobs like this, the Argon, Boron, and Xenon microcontrollers from Particle all provide built-in mesh-networking capabilities.

For this build, I used the Particle Mesh platform to create a network of R/C cars, each controlled by a Particle Xenon. All of the Xenon R/C cars are endpoints, and are connected to a mesh network with a single Particle Argon as the gateway. Once the network is established, I can use local network messaging to send low-latency commands to all nodes on the network and make my R/C cars dance.

First, however, I needed to hack the off-the-shelf Tumblers to Particle-power them!

1. HACK THE THUNDER TUMBLER

The first step is hacking the R/C car to add a Xenon. Whichever brand or style of R/C car you're using, the objective is to crack open the car to reveal the PCB inside, determine how the device sends commands to the motors to rotate the wheels, and then connect pins from the Xenon to the corresponding motor driver pins on the car. Because these cars are inexpensive, you should expect to find some variation, even among those with the Thunder Tumbler name on the box.

Write-ups on Tumbler hacking can be found online going back over 9 years. That's a lifetime in the electronics world, so be sure to test and verify the functionality of your cars as you follow these instructions. Grab your favorite multimeter and measure voltages across various pins on the R/C car's PCB as you make the wheels spin with the remote control. Make sure to set your car somewhere where the wheels can rotate freely so that it doesn't get away from you during testing (Figure A).

Remove the two screws that hold the car body in place, then lift the shell to expose the PCB. You'll see 8–10 wires that run from the car to the PCB, and a number of through-hole and surface-mount components (Figure B).

The small surface-mount component on the top is a radio-control receiver chip, the RX-2B. Its pair, the TX-2B transceiver chip, sits in the remote control transmitter. These ICs are commonly used for R/C vehicles,

and I was able to find their datasheets to determine which pins map to the forward, backward, left, and right commands from the remote. I also used the datasheet to establish a known ground pin; you'll need this to determine which pins power the car motors and can be controlled by a Xenon.

This ended up being critical because the main IC on these boards, through-hole mounted on the bottom of the car PCB, is one for which I couldn't find a datasheet, in spite of many hours of searching. While I'm still not 100% certain of all the features on this chip, it functions primarily as an H-bridge motor controller. Pulsing a signal into a certain pin on the IC results in a pulse out to one of the motor control pins, which makes a wheel spin forward or backward.

By reverse engineering, I found that there are four pins I care about on this unknown IC: one that spins the left wheel forward, one that spins it back; one that spins the right wheel forward, and one that spins it back. I soldered one wire to the top of each of these pins, and one to a ground pin (Figure C). Even though the Xenon and car are powered separately, they need to share a ground for everything to work right.

Then I connected these wires to Xenon pins A1, A0, A2, A3, and GND, and powered up the Xenon with a LiPo battery (Figure D).

2. SET UP A MESH NETWORK

Your mesh will have one gateway (a Particle Argon or Boron) and one Xenon-based node for each R/C car. If you want to program your Xenons over-the-air without having to connect each to a computer, you'll want to set up your network in advance. You can do this from the Particle mobile app (Figure E, on the following pages), or follow the instructions at particle.io/start.

3. PROGRAM THE TUMBLERS

Once you have your first Xenon claimed and ready, the next step is to write firmware to the Xenon for controlling the actions of each car. Since the goal here is to create a small mesh of swarmbots, you'll start with a simple test sequence that moves the car forward, back, left, and right. The code for this sequence can be found on the following page and at github.com/particle-iot/mesh-rc-cars. Upload it to your Xenon just as you would to an Arduino; if you need instructions they're in a Readme at the GitHub repo.

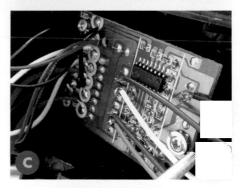

BRANDON SATROM is developer advocate for Particle and founder of Carrot Pants Studios, a maker education company. An avid tinkerer, he loves to talk sensors and circuits, microcontrollers, open source, robots, and other new shiny tools and technologies.

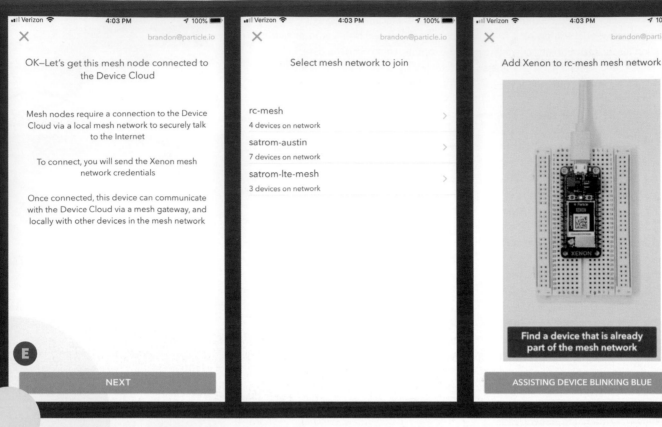

```
  // Wheel pin mappings
  int leftReverse = A0;
int leftForward = A1;
int rightForward = A2;
int rightReverse = A3;

// Speed and delay variables
int speed = 85;
int turnSpeed = 255;
int forwardDelay = 1000;
int backDelay = 1000;
int turnDelay = 2000;

void setup()
{
  // Set motor pins to outputs
  pinMode(leftReverse, OUTPUT);
  pinMode(leftForward, OUTPUT);
  pinMode(rightForward, OUTPUT);
  pinMode(rightReverse, OUTPUT);

  // Make sure each motor is off
  digitalWrite(leftReverse, LOW);
  digitalWrite(leftForward, LOW);
  digitalWrite(rightForward,
LOW);
  digitalWrite(rightReverse,
LOW);
```

```
}

void runDemo(const char *event,
const char *data)
{
  allOff();

  goForward(speed);
  delay(forwardDelay);

  goBack(speed);
  delay(backDelay);

  // Max spin to raise up on the
back tires
  turnLeft(turnSpeed);
  delay(turnDelay);

  allOff();
}

void allOff()
{
  analogWrite(leftReverse, 0);
  analogWrite(leftForward, 0);
  analogWrite(rightForward, 0);
  analogWrite(rightReverse, 0);
```

```
  delay(50);
}

void goForward(int speed)
{
  allOff();

  analogWrite(rightForward,
speed);
  analogWrite(leftForward,
speed);
}

void goBack(int speed)
{
  allOff();

  analogWrite(rightReverse,
speed);
  analogWrite(leftReverse,
speed);
}

void turnLeft(int speed)
{
  allOff();

  analogWrite(rightForward,
```

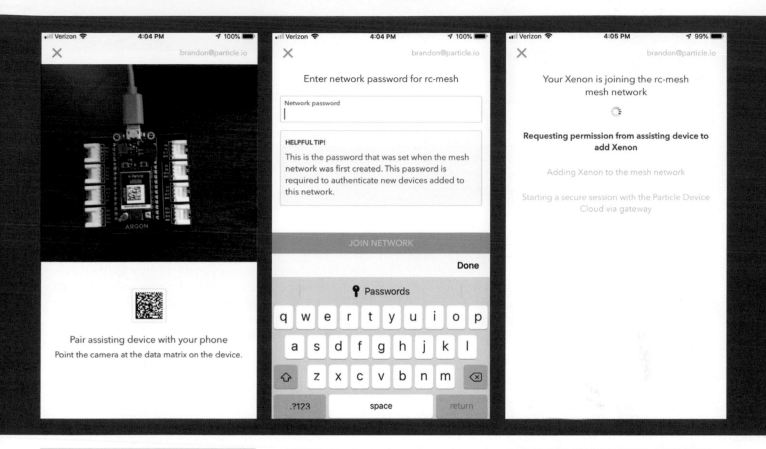

Pair assisting device with your phone
Point the camera at the data matrix on the device.

```
 speed);
}

void turnRight(int speed)
{
  allOff();

  analogWrite(leftForward,
speed);
}

void loop()
{
  // Nothing needed here.
}
```

You probably noticed that the meat of this demo is **analogWrite** commands to turn each wheel forward or backward by sending a voltage to a corresponding pin. Through testing, I determined that the cars I'm using are PWM-able, meaning that I can set a wheel pin to lower voltages than digital **HIGH** (3.3 volts in the case of the Xenon) and have the wheel turn at lower speeds. Notice that I'm passing in different values for turning right or left versus forward and back. PWM (pulse-width modulation) enables pretty complex patterns and speeds for these inexpensive cars!

With the demo code done, the final piece you need on your R/C cars is a subscription to a local network message that specifies the **runDemo** function as the handler. Particle enables local, in-mesh-network messaging between nodes on a network using a **Mesh.publish** and **Mesh.subscribe** API. **Mesh.publish** sends a broadcast message (with a name and payload) to all nodes on the network. **Mesh.subscribe**, on the other hand, listens for messages (by name) and specifies a handler to process and respond to those incoming messages.

Our Xenons will use **Mesh.subscribe** to listen for a network message, and then trigger the demo sequence.

```
// Add to setup function
Mesh.subscribe("run-demo",
runDemo);
```

Flash the completed code to each R/C car Xenon and you're ready for the final piece: programming the gateway to coordinate the movements of the car swarm.

4. PROGRAM THE MESH GATEWAY

The gateway code that runs on the Argon is less complex than the code on the Xenons, but still needs two pieces: a **Mesh.publish** call to trigger the R/C car demo sequence, and a way to tell the Argon to fire that message out to the mesh network.

You could just fire the message when the gateway comes online, or regularly on a delay, but what fun would that be? Instead, you'll use another Particle API, **Particle.function**, to trigger the cars from a mobile phone whenever you want.

```
void setup() {
  Particle.function("runDemo",
runDemo);
}

int runDemo (String command)
{
  Mesh.publish("run-demo", NULL);

  return 1;
}

void loop() {
  // Nothing needed here
}
```

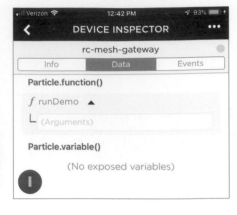

Particle.function takes a string value name of the function (used when calling that function from a mobile or web app or through the cloud) and a handler function to execute when called. That function contains our **Mesh. publish** call, which sends a broadcast message with the name "run-demo" to all listeners (i.e. our R/C cars) on the network. Once you've added the code on the previous page to your gateway, you're ready to control some swarmbots (Figure **F**).

5. PUT IT ALL TOGETHER

Particle functions can be called from any device that has a secure connection to the Particle Device Cloud, which could be the browser-based Particle Console at console.particle.io (Figure **G**), the Particle command-line interface or CLI (Figure **H**), or your own applications. They can also be called from the Particle mobile app (Figure **I**), which seems like a fitting place to finish this project.

The mobile app shows a listing of every Particle device you own. Click on your gateway, then the Data tab menu, and you'll see a "runDemo" function. Put your Xenons in position with plenty of space, and click the function to trigger some R/C swarming!

The demo code illustrates a simple, coordinated sequence of actions. But with PWM and some trial and error, you can create any number of cool synchronized dances with mesh-networked R/C cars.

COMMAND YOUR SWARM!

Now that your bots are meshed, let's turn them into real swarmbots that can communicate with each other. A common swarm scenario is the leader-follower pattern. The leader determines movements for the swarm, and communicates these directly to all followers (Figure **J** and **K**). One example would be to control the leader with the out-of-box remote control, read pin voltages off the leader car, and send these as instructions to all followers listening for a **Mesh** event.

To build this demo, I designated one of my three cars as a leader, and added firmware to control the fleet. Instead of writing to my motor pins, the leader simply reads the analog values sent by the remote control and sends these to listeners on the network with a **Mesh.publish** that includes the wheel, direction, and analog value to apply to follower cars.

```
int32_t lastLeftRVal = 0;
int32_t lastLeftFVal = 0;
int32_t lastRightRVal = 0;
int32_t lastRightFVal = 0;

#define MIN_PIN_VAL 150
#define DRIVE_VAL 200

void setup()
{
  pinMode(leftReverse, INPUT);
  pinMode(leftForward, INPUT);
  pinMode(rightForward, INPUT);
  pinMode(rightReverse, INPUT);
}

void loop()
{
  checkPin(leftReverse,
&lastLeftRVal, "leftR");
  checkPin(leftForward,
&lastLeftFVal, "leftF");
  checkPin(rightReverse,
&lastRightRVal, "rightR");
  checkPin(rightForward,
&lastRightFVal, "rightF");
}

void checkPin(int pin, int32_t
*lastVal, const char *event)
{
  int32_t pinVal =
analogRead(pin) / 16;

  if (pinVal > MIN_PIN_VAL)
    pinVal = DRIVE_VAL;
  else
    pinVal = 0;

  if (pinVal != *lastVal &&
pinVal == DRIVE_VAL)
  {
    *lastVal = pinVal;

    Mesh.publish(event,
String(DRIVE_VAL));
  } else if (pinVal == 0 &&
*lastVal != 0) {
    *lastVal = 0;

    Mesh.publish(event,
String(0));
  }
}
```

On the follower side, I added firmware that listens for a **Mesh** event for each pin, and performs an **analogWrite** with the passed-in value. On the cars themselves, I also removed the antenna on each board and cut the traces from the remote's RX IC just to make sure that movement commands were only coming from the leader.

```
void setup()
{
  pinMode(leftReverse, OUTPUT);
  pinMode(leftForward, OUTPUT);
  pinMode(rightForward, OUTPUT);
  pinMode(rightReverse, OUTPUT);

  digitalWrite(leftReverse, LOW);
  digitalWrite(leftForward, LOW);
  digitalWrite(rightForward,
LOW);
  digitalWrite(rightReverse,
LOW);

  Mesh.subscribe("leftR", leftR);
  Mesh.subscribe("leftF", leftF);
  Mesh.subscribe("rightR",
rightR);
  Mesh.subscribe("rightF",
rightF);
}

void leftR(const char *event,
const char *data)
{
  move(leftReverse, data);
}

void leftF(const char *event,
const char *data)
{
  move(leftForward, data);
}

void rightR(const char *event,
const char *data)
{
  move(rightReverse, data);
}

void rightF(const char *event,
const char *data)
{
  move(rightForward, data);
}
```

```
void move(int pin, const char
*speed)
{
  int32_t speedVal = atoi(speed);

  if (speedVal > 16) // Filter
out noise from the leader
  {
    analogWrite(pin, speedVal);
  }
  else
  {
    analogWrite(pin, 0);
  }
}

void loop() {}
```

With this simple R/C-controlled swarm as our foundation, we built a bunch more swarm sequences:

» **Follow the leader** — basic forward and back demo with ending spin
» **Splinter** — cars separate in three different directions and come back together
» **Follow the leader and push** — leader tells followers to stop, goes forward 2 seconds, turns around, goes back 2 seconds, then tells followers to move backward as it keeps moving forward
» **Sentry mode** — square path with right-angle turns
» **Orbit** — followers orbit a stationary leader.
  You can grab the complete project source code, including these more complex demos, from the repository at github.com/particle-iot/mesh-rc-cars. I've also done a number of Twitch streams focused on building the demos for this project; check them out at twitch.tv/brandonsatrom.

HACK IT AND SHARE IT

We'd love to see makers extend and modify this project with sensors and new behaviors. You could add collision detection to the leader with a PIR or ultrasonic sensor, or even use the remote to log predefined sequences that could then be repeated automatically. And if you've built your own mesh-style swarmbot project, or anything else with Particle's new hardware, we'd love to hear from you! Share with us on Twitter (@Particle), Instagram (@Particle_io) or in our community of 160,000 developers at community.particle.io.
  Swarm on, makers! ⊙

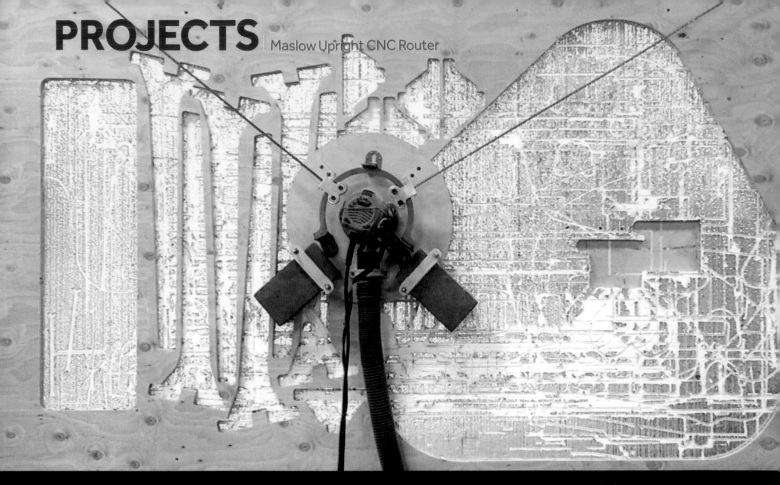

Make a **Maslow CNC**

Build the unique, upright CNC router that cuts full 4'×8' sheets but takes little floor space **Written by Bar Smith, Chris Skiles, and Patrick Kinnamon**

BAR SMITH studied electrical engineering at the University of California, Santa Cruz where he fell in love with digital design and fabrication. He has run two Kickstarter campaigns for open source CNC routers (Makesmith and Maslow) and is now working on CAD/CAM software for upright CNC machines.

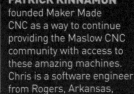

CHRIS SKILES and **PATRICK KINNAMON** founded Maker Made CNC as a way to continue providing the Maslow CNC community with access to these amazing machines. Chris is a software engineer from Rogers, Arkansas, who enjoys making web and mobile apps and building things with his wife and five kids. Patrick is a Texas native and New York resident with a background in arts and analytics, and a passion for bringing creative solutions to market.

The Maslow is a large format CNC router capable of cutting wood and other materials with precision and repeatability, based on a digital file. A CNC (computer numerical control) machine allows makers to automate the cutting process in woodworking and other manufacturing operations. When the Maslow kit is built and software is installed, makers will have a fully functional CNC machine able to cut a surface of 4×8 feet, with adjustable cut depth. The design is open source so anyone can build it, from scratch or from a kit.

The application for this tool is enormous. Cabinet makers, hobbyists, sign makers, woodworkers, furniture makers, and more have all seen the value in automated cutting that can multiply their productivity and produce extremely intricate designs. Whether for personal use, side hustle, or full-time day job, the Maslow CNC is an incredibly versatile tool. You can build it for

about $600 if you've already got an ordinary fixed-base router, and unlike traditional flatbed CNCs, its upright design takes up very little floor space in your shop or garage.

Maslow began as a hobby project of Bar Smith in 2015. In 2016 Hannah Teagle joined to help run a Kickstarter campaign to build an open source community around the project. In 2017 they showed the machine at Maker Faire Bay Area, shipped four batches of kits sending thousands of vertical CNC machines around the world, and watched the community take off.

Soon the logistics of packing and shipping kits became too much, and Bar and Hannah asked the Maslow community for help. They publicly posted the design files and made the factory tooling available to anyone who was willing to take on the task of continuing to make kits.

Maker Made CNC stepped forward and made kits available again only a few months

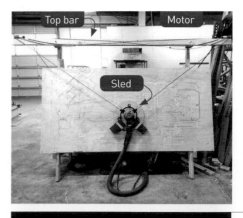

TIME REQUIRED:
A Weekend

DIFFICULTY:
Advanced

COST:
$600–$750

MATERIALS

ROUTER
- » **Router, fixed base, with depth adjustment** We recommend the Ridgid R22002 for U.S. customers. International users have been successful with Bosch routers.

FRAME
- » ¾" plywood, 4'×8' sheet
- » 2×4 lumber: 10' lengths (6) and 8' lengths (2)
- » Bricks (2)
- » Wood glue

You can build your Maslow from a kit by Maker Made CNC ($499, makermadecnc.com), and they're offering free shipping to *Make:* readers (use coupon code MakeMag2019). Or you can build it from scratch by sourcing the parts yourself. Key components include:

ELECTRONICS
- » **Arduino Mega 2560 microcontroller board** Maker Made (MM) part #810 or Amazon #B01H4ZLZLQ
- » **CNC motor controller shield for Arduino** MM #800, or maslowsurplusparts.com/products/arduino-shield-v1-2. Or make your own, using the design files and components list at github.com/MaslowCNC/Electronics.
- » **Heat sinks (2)** included in MM #800, or try Digi-Key #345-1102-ND
- » **Gearmotors, worm drive, 191:1 reduction, 7ppr encoder (2)** MM #880, or maslowsurplusparts.com/products/maslow-cnc-motor-single. The key is that the encoder is on the *back side* of the gearbox so the 7ppr encoder gets scaled up 191x, then quadrature encoding gives us 4 steps per pulse, so at the output shaft you have 7*191*4 = 5,348 steps per rotation.
- » **Motor cables, custom (2)** MM #900
- » **Power supply, 12V 5A, 2.5mm × 5.5mm plug** MM #850 or Amazon #B07D3TCXVK
- » **Z-axis gearmotor kit with cable, mounting brackets, and shaft coupler** MM #910

BRACKETS
- » **Motor mounts (2)** for the X-Y motors; MM #890 or Amazon #B073NZ81M9
- » **Router sled ring, custom** included in the Maker Made ring kit, #860. The CAD files for these parts are at github.com/MaslowCNC, so your local laser cutting or waterjet company could fabricate them easily.
- » **Sled ring bearing carriage brackets (2)** included in MM #860
- » **L-brackets (5)** included in MM #910 and 860, or try maslowsurplusparts.com/products/l-bracket

OTHER
- » **Sled ring bearings (4)** included in MM #860, or maslowsurplusparts.com/products/sled-ring-bearings
- » **Roller chains, #25 (2)** MM# 830 or Amazon #B018H9ZAD2
- » **Motor sprockets for #25 chain, 10 tooth, 8mm bore (2)** included in MM hardware kit #840; or the 9-tooth #SPR-2509P from electricscooterparts.com. There's software setting for the number of teeth.

CONTINUED ON PAGE 44 ...

after the last kit from Maslow shipped. Their mission is to continue the work of Bar and others by investing in updates to make the kit accessible to a wider audience of makers, and introduce products and accessories that will further enhance its functionality and capabilities.

BUILDING YOUR MASLOW CNC
Building the Maslow is a true DIY project. You can follow the complete, user-editable assembly guide at maslowcnc.com/assemblyguide.

Scratch builders can also get schematics, blueprints, and CAD files for the custom parts on the Maslow website, so a local shop with a waterjet cutter and a press brake could easily make them.

Here's an overview of how it all goes together.

1. BUILD THE FRAME
To reduce cost and ensure the Maslow is available to as many makers as possible, the frame is built by the user from materials sourced at their local hardware store. While there are multiple options for the frame, we recommend the wooden Default Frame designed by the Maslow community (Figure Ⓐ); it's the most common and the simplest to construct . Full plans can be found at maslowcommunitygarden.org/The-Default-Frame.html.

The top crossbeam, which holds the X and Y motors in place, is the most structurally critical element in the frame. This piece is made from a single, 10-foot-long 2×4 beam. The rest of the frame is there to support the piece of plywood being cut. The default design is made from eight 2×4s screwed and glued together.

Frame customization is also common, with designs made to fit user spaces. Some fold away when not in use, are enclosed to reduce noise, or make the machine larger or smaller than the default design.

Once the frame is built, you'll mount the X-Y motors and tensioning pulleys, and hang the drive chains.

2. ASSEMBLE THE RING
At the center of the machine is a ring where the two drive chains meet to support the router. This ring kit (Figure Ⓑ) ensures that as the sled shifts and rotates from the pull of the motors, the router bit remains centered and undisturbed for a smooth cut

WHAT WILL YOU CUT?
Your Maslow CNC can cut anything that can be cut with a hand router. Maximum thickness is only restricted by your bit size and router travel. The recommended Ridgid R22002 router has 1½" of travel, and the machine will make multiple shallow passes to cut through thicker materials. The most popular materials are:
- » **Hardwoods**
- » **Plywood**
- » **MDF**
- » **Laminate**
- » **Hard plastics**: acrylic, nylon, uPVC, hard PVC
- » **Soft plastics**: polycarbonate, polyethylene, soft PVC

MATERIALS continued

» **Slack sprockets for #25 chain, 16 tooth (2)** included in MM #840, or robotshop.com #RB-Sct-228 looks like a good alternative.
» **Pulleys, 4mm × 13mm × 7mm (2)** included in MM #840, or use Amazon #B017691AJW
» **Elastic cords (2)** MM #820 or Amazon #B06W5279YL
» **Assorted bolts, nuts, washers, screws, and fasteners** included in MM #840; see complete bill of materials at maslowcommunitygarden.org/Maslow-CNC-Kit.html?buy=true.

TOOLS

» **Screwdriver or drill/driver**
» **Wood saw**
» **Pliers**
» **Wood glue**
» **Computer with Arduino IDE and Ground Control software** free downloads at arduino.cc/downloads and github.com/MaslowCNC/GroundControl

Connecting cables to motors.

Connecting motor cables to the shield.

Connecting the Arduino to the motor shield.

Electronic components assembled.

line. Full instructions for assembly can be found at maslowcommunitygarden.org/Maslow-Ring-System.html.

3. ASSEMBLE THE ELECTRONICS

Maslow is built around time-tested electronics. An Arduino Mega 2560 is used as the "brains" of the machine, while the custom motor controller shield allows for individual control of the multiple motors (Figure C). The motor controller plugs into the Maslow's three gearmotors (we use a closed-loop control system which is a little fancier than a stepper motor) and connects via USB to a computer where the machine is controlled using software openly available for Windows, Mac, or Linux machines.

No soldering is required to put the electronics together, although the open architecture can be modified to add additional features.

4. SET UP THE SOFTWARE

To run the tool, you need two pieces of software: the machine control software for your computer (Figure D), and the firmware for the Arduino. Both are free (and both come pre-loaded on a flash drive that comes with your kit from Maker Made).

Download the open source Maslow Firmware from github.com/MaslowCNC/Firmware, then follow the instructions there to upload it to the Arduino Mega.

The machine is controlled using open source software called Ground Control, available for Windows, Mac, or Linux computers (Figure E). From within Ground Control, you can move the machine to where you want to begin a cut, calibrate it, open and run a G-code file, or monitor the progress of an ongoing cut.

Ground Control is written in Python because it has good cross platform support and is relatively simple to work with. Two of the goals of Ground Control are: for it to run on as many platforms as possible, and to be as easy as possible for the community to contribute to making the program better. You can download Ground Control at github.com/MaslowCNC/GroundControl/releases. Follow the instructions at maslowcommunitygarden.org/GroundControl.html to install it, then check out our User Guide at github.com/MaslowCNC/GroundControl/wiki/Ground-Control-Users-Guide to learn how to run it.

The machine uses G-code, which is a very popular file type among CNC users and hundreds of software applications. These applications will accept varying design file types, with DXF and SVG being the most popular. DXF files are commonly created using AutoCAD, and SVG files can be created using design programs like Illustrator, Inkscape, and Gimp.

5. CUT THE FINAL SLED

To function properly, the Maslow kit requires one component — the round sled — to be cut by CNC. Of course, this leads to a "chicken or the egg" question: How can the machine cut parts for itself before it is built?

The solution is to manually cut and install a rough, square version of the sled (seen in Figure B on the previous page), at which point the machine is complete enough to cut a replacement sled. Cutting the final sled with all its mounting holes is a great first project and a useful way to learn to use the machine. A successful build (Figures F and G) will guarantee that the router's bit rests at the very center point of the sled, ensuring that the bit will remain along its designated cut line, regardless of how the sled may rotate.

With the final sled mounted, your Maslow is ready to cut (Figure H).

6. ADD THE Z-AXIS

Once you've got your final sled working, go ahead and mount the Z-axis motor and brackets, then couple the motor shaft to your router's depth adjustment screw (Figures I, J, and K). Plug the motor cable into the Arduino motor shield, enable the Z-axis in the Ground Control software, and you're ready to start using your automated Z-axis (Figure L)!

GOING FURTHER

The Maslow community has made countless improvements to the basic design and also shared many modifications. One of the most popular upgrades is the Meticulous Z-Axis (Figure M), with a plywood gantry and leadscrew that move the entire router (not just the depth adjustment), a plexiglass dust window, 3D-printed dust chute, and more. Check it out at maslowcommunitygarden.org/The-Meticulous-Z-Axis.html and click on the Forums tab to see the 200+ post

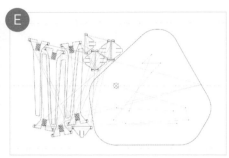

Setting up a cutting job on the Ground Control CNC software.

Mounting the sled.

The final sled with ring kit.

Maker Made CNC founders Chris Skiles (left) and Patrick Kinnamon (right) show their completed Maslow CNC kit, ready to cut.

conversation that led to that design. Another good thread is: forums.maslowcnc.com/t/z-axis-upgrades-consolidation/5392.

We hope this brief introduction demonstrates the amazing possibilities of the Maslow CNC and gives a flavor of the build process. One of the great advantages of the Maslow kit is the incredible dedication of its community of supporters. To really become a Maslow master maker, spend some time at forums.maslowcnc.com. You'll find an enormous wealth of support, project ideas, tips, tricks, giveaways, and discussions that will take your skills to another level. ✪

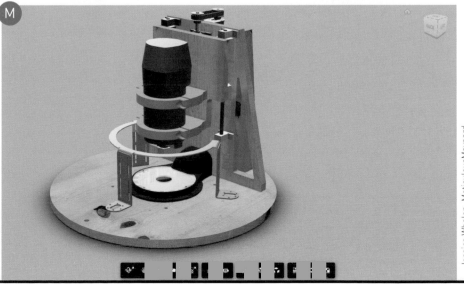

Jessica Whalen, MeticulousMaynard

CELL COMPUTER CASE

by Markus Simon

Markus Simon in Seewalchen, Austria, (username simon_makes) built this really cool computer case and did a series of YouTube videos about the process. I like this one because those parts would be really tough to cut by hand, *and* because he actually messed up and ruined the first one he made, but the CNC let him cut out a new one really easily.

forums.maslowcnc.com/t/cell-pc-case-build/5799/10

RED WAGON FIGHTER JET

by Grant and Kenzie Stucki, and Jakrro

Grant Stucki (username GMS10) in College Station, Texas, posted that his wife Kenzie wanted to turn their kids' red wagon into a fighter jet for Halloween. He shared dimensions and asked if anyone in the community could help with the CAD. Right away community member Jakrro in Gjøvik, Norway, jumped in to help; he used SolidWorks to draw up plans that Grant and Kenzie then cut out and painted for an adorable family Halloween costume.

maslowcommunitygarden.org/Red-Wagon-Fighter-Jet.html

Vert Tricks! Written by Bar Smith

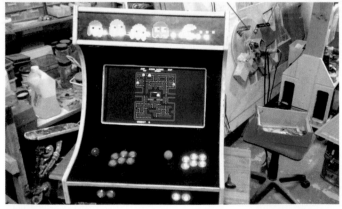

ARCADE CABINETS

by KingOfScolboa

Community member KingOfScolboa in Parkgate, Northern Ireland, posted some amazing open source Raspberry Pi arcade cabinet plans at maslowcommunitygarden.org/Tabletop-Arcade.html. It's a cool project with a lot of appeal — we're seeing other people making them too now, like a batch of four from Alex Benz (username Delloman) at the Maker Station in Marietta, Georgia.

forums.maslowcnc.com/t/arcade-cabinet-plans/1669/29
forums.maslowcnc.com/t/maslow-to-cut-an-arcade-cabinet

OUTDOOR SIGN FOR COFFEE/BIKE SHOP

by Mason Adkins

We see a lot of signs being made — I like this one just because it's so big! This LED-lit outdoor sign was cut from cellular PVC boards from Home Depot, by Mason Adkins (username Jeebis), co-owner of Sixth City Cycles in Cleveland, Ohio. He set up his Maslow in the basement of the bike shop.

forums.maslowcnc.com/t/outdoor-sign-for-coffee-bicycle-shop

BOLGER BOBCAT SAILBOAT
by Sonny Lacey

Sonny in Baltimore, Maryland, built a sailboat, which is pretty fantastic! He bought blueprints for a stitch-and-glue build, and digitized them in Fusion 360 to cut on his Maslow. He reports: "The *Manx*, as she is now known, is laying peaceably in a slip at Anchorage Marina in Baltimore's Inner Harbor." This design is copyrighted, but I think it's a matter of time before we see open source boat designs.
forums.maslowcnc.com/t/building-a-bolger-bobcat-payson-tiny-cat-catboat-from-cad-on-up

OPENDESK FURNITURE
by various users

Another category of project we see often is open source furniture designs from Opendesk (opendesk.cc). Here's a Slim Chair that Mark den Otter in the Netherlands (username MdO) scaled down to be the right size for his kids — with no prior experience in CNC or CAD. The Nimble Stool design is also popular.
forums.maslowcnc.com/t/slim-chair-from-opendesk
forums.maslowcnc.com/t/opendesk-nimble-stool

Makers are creating amazing projects with their **Maslow vertical CNC routers**. Here are a few of my favorites from the Maslow community.

OPEN SOURCE BAT HOUSES
by Didier Destabeaux

Community member Didier designed an open source bat house using his Maslow, then had hundreds of kits made by a local subcontractor (because Maslow uses all the normal file formats, it was easy to do). Now his nonprofit *Tous aux abris!* is distributing them throughout France to help restore bat populations in decline. A former prime minister even dropped by to see his Maslow in action.
forums.maslowcnc.com/t/a-christmas-tale-for-maslownians

PEDAL TRUCK
by James Coffey

James (username Sreenigne), an engineer in Pembroke, Georgia, made this awesome pedal car which I think is just an excellent bit of craftsmanship. This Roadster/Pie Truck hybrid is his own modified version of DIY plans from Stevenson Projects.
forums.maslowcnc.com/t/3-way-pedal-truck-project-stevenson-project

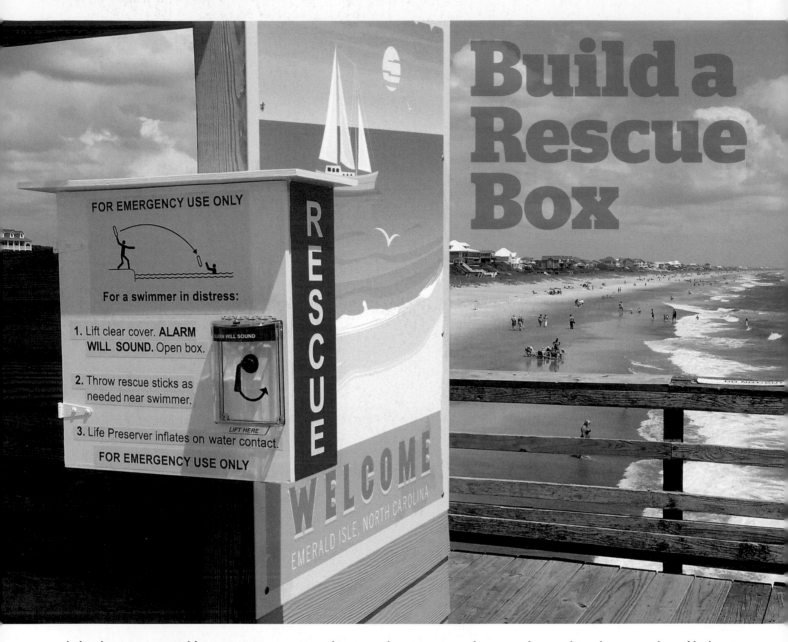

Build a Rescue Box

FOR EMERGENCY USE ONLY

For a swimmer in distress:

1. Lift clear cover. **ALARM WILL SOUND.** Open box.

2. Throw rescue sticks as needed near swimmer.

3. Life Preserver inflates on water contact.

FOR EMERGENCY USE ONLY

ALARM WILL SOUND

LIFT HERE

RESCUE

WELCOME

EMERALD ISLE, NORTH CAROLINA

Help save lives at your beach or swimming hole — build a tamper-proof box for throwable "Rescue Sticks" Written by Phil Bowie

PHIL BOWIE is a pilot, motorcycle rider, fiddler, inventor, and Coast Guard–licensed boat captain. He lives by the wide Neuse River in eastern North Carolina. He's the author of four acclaimed suspense novels in a series available in print and e-book on Amazon. Visit him at philbowie.com.

A North Carolina lifeguard told me, "A drowning happens fast. A struggling, inexperienced swimmer has only a minute or two before it's too late."

Lifeguards can beat that fast-ticking clock, and they routinely do every season hundreds of times across the country. But in swimming areas without lifeguards there is often no realistic chance of rescue. According to the CDC, there are 3,500 non-boating-related drowning deaths in America every year, or about 10 per day — the fifth leading cause of unintended injury death.

Bystanders should always call 911, of course, but many times rescue personnel, hard as they try, cannot get to the scene fast enough. And it's never a good idea for an untrained bystander to attempt a swimming rescue because statistics show that can easily turn into a double drowning.

Last summer at an ocean beach near where I live, a woman in her forties was suddenly caught in a rip current. All bystanders could do was watch her die. The recorded 911 call was hard to listen to, and right then I vowed to devise a rescue system that could be constantly available at any swimming area and could rival the fast response time of a lifeguard.

The result is the Rescue Box. It's alarmed against tampering, like a fire alarm, but is quickly and easily accessible by anyone in a potential drowning emergency. It houses three Mustang Rescue Sticks (Figures A and B) which can be thrown — even into a wind — 100 feet or more to reach a distressed swimmer. The foam-padded 15.5oz Stick inflates immediately on water contact, creating a large, bright-yellow, horseshoe-shaped preserver with twice the flotation of a traditional life jacket, which cannot be thrown very far at all.

Mustang Sticks are proven lifesavers for boaters and first responders, but the Rescue Box is a n... making the Sticks available t... and whenever n... ever thrown a s... snowballs, ca... enough to sa... box holds th... chances to...

The Res... project fo... beach or... like the... college... taking... civic... a loc... swi... kn...

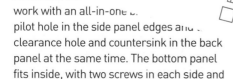
screws for e...
work with an all-in-one b...
pilot hole in the side panel edges and s...
clearance hole and countersink in the back panel at the same time. The bottom panel fits inside, with two screws in each side and

TIME REQUIRED:
A Weekend or Two

DIFFICULTY:
Intermediate

COST:
$250–$275 plus Rescue Sticks

MATERIALS

» **Mustang Survival Rescue Sticks (3)** #MRD100. E-mail Alex Moret at amoret@ mustangsurvival.com for special program pricing.
» **Plywood, ½" thick, good both sides, 4'×4'** I used Lowe's "project plywood" but I recommend "marine" grade if you can find it. Cut 7 pieces for the box:

   Back — 20" wide × 21" high
   Sides (2) — 6"×21"
   Top — 9"×24"
   Bottom — 6"×19"
   Top filler strip — 1"×19"
   Door — 20" wide × 20⅞" high

» ...hinges, 3", brass or stainless (2)
» ...per, alarmed, flush ...Safety item ...ours ...ARM

...Slide-

...p, 5/16"
...0 Building

...white semi- ...Extreme ...lent

...t orange paint ...d spray paint to

» Clear silicone sealant
» Aluminum strip, ½"×⅛"×48"
» Aluminum angle, ¾"×¾"×⅛"×38"
» Triangular bicycle flag, day-glow orange or red with fiberglass wand
» Lightweight chain, 12"
» Vinyl graphics
» Wood filler or body putty

NOTE: All hardware listed below should be stainless steel.

» Wood screws, flat head: #6×1" and #6×½"
» Wood screws: round or pan head: #6×½" and #8×½"
» Machine screws, #8×¾"
» Plain washers, #6 and #8
» Locknuts, #8
» Hanger bolts, ¼-20 × 2½" (2)
» Washers, ¼" ID × 1½" OD (2)
» Washers, ¼" ID × 1" OD (2)
» Lock washers, ¼" (2)
» Nuts, ¼-20 (2)

TOOLS
» **Table saw**
» **Band saw or hacksaw**
» **Drill press or hand drill**
» **Drill bits, various** It's nice to have an "all in one" pilot/clearance/countersink bit for the 1" #6 flat head screws.
» **Phillips screwdriver or driver bit**
» **Wrenches** for #8 nuts and ¼-20 nuts
» **Sandpaper, 80 and 120 grit**
» **Metal file**
» **Hammer**
» **Center punch and lubricating oil** for drilling holes in aluminum
» **Steel rule**
» **Paintbrush**
» **Bench vise**
» **Utility knife** for mitering weather seal
» **Pencil and fine point felt-tip marker**
» **Tap** for 6-32 machine screw

...ll the top panel flush ...nging on the front and ...oor from weather, with ...side and three into the ...ont filler strip with three ...ugh the top panel.

...AINT
...ad depressions and any ...wood edges with wood filler ...utty and sand smooth. Knock ...es with 120 grit sandpaper. ...num durability, paint the box ...side and out with a coat of white ...owed by two coats of exterior se...s acrylic latex, sanding lightly between coats if necessary to keep the finish smooth.

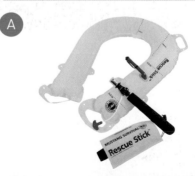
A

B

3. MAKE THE STICK HANGERS

Make the Rescue Stick hangers from six ½"×⅛"×7¼" aluminum strips (Figure ). The bottom of the hanger is bent down 90°, 1½" from the end, and the top is bent up 60°, 1½" from that end. To create each bend, clamp the strip in a bench vise close to the bend point and tap with a hammer until the desired angle is achieved. Use your first hanger as a template for the other five. Drill two screw clearance holes in each lower end. Use a file and coarse sandpaper to round and smooth the cut ends so they will not injure the user or snag the Stick fabric.

With ½" #6 round head wood screws and optional washers, mount the hangers to the inside box back, 9" apart on center (OC), each pair centered sideways in the box (Figure ). The Stick-resting surfaces of the top pair should be 5¾" down from the top, with the other pairs of resting surfaces each 5½" down from the pair above. These dimensions aren't critical; what's important is that the Sticks are easy to grab.

4. WEATHERPROOF THE BOX

Attach the rubber weatherstrip to the front box edges and the top filler strip. Reinforce the self-stick adhesive with a single dot of gel super glue every 2"–3" and at all four mitered corners. Fill any corner gaps with clear silicone for a good weather-tight door seal all around.

5. MOUNT THE DOOR

Install the two brass or stainless strap hinges on the door, pin side out, using ¾" #8 machine screws, plain washers, and locknuts. Because a hole or two may be too close to the hinge pin, you may have to drill new holes so you have at least two screws in each hinge wing. Fill any unused holes with body putty and sand smooth. I painted the hinges for better appearance. The hinges should be about 4" OC from the top and bottom of the door. Lay the box flat on its back. Rest the door on the weatherstripping, which will compress just the right amount from the door weight alone. Mark the hinge holes for drilling through the box side, and install the door with ¾" #8 machine screws, washers, and locknuts.

Make a vertical door stiffener from ¾"×¾"×⅛" aluminum angle, 17" long. Install it 1¾" down from the door top and 4⅞" in from the door edge, using three ½" #8 round head wood screws. This

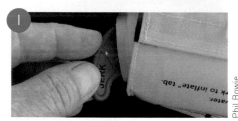

Phil Bowie

will prevent warping and also serve as a door latch mechanism stop. To keep the bottom panel flat, you can optionally add an aluminum angle 18¾" long, mounted with three ½" pan-head wood screws so it's flush with the front edge or about 1" back. You can see both stiffeners in Figure B.

Install the chain door retainer (Figure ) with two ½" #8 round head wood screws.

6. INSTALL THE "STOPPER" ALARM

The gasketed clear alarmed door handle cover (Figure F) is called a *stopper*, meaning it's intended to stop tampering, which it does very well. It comes with directions. You can have a custom message printed on it when ordering. The alarm is

powered by a 9V battery; it sounds loudly when the cover is lifted and stops when the cover is closed. It comes with a pin and fob in place to keep it silent until installation is finished (Figure F shows the fob).

Install the stopper 4¼" up from the door bottom and ⅜" in from the right-hand edge, using four ½" #6 flat head wood screws.

7. MOUNT THE DOOR LATCH

I used the L-handle — painted red to match the alarm box — and the threaded pin from a brass storm-door lock set to make the latch mechanism (Figure G). Drill a clearance hole through the door 8⅛" up from the bottom door edge and 3½" in from the right edge for the threaded knob-set

The latch catch (Figure **H**) is made from a 2" length of ¾"×¾"×⅛" aluminum angle, and has a smooth filed chamfer on the underside top lip to allow easy gradual engagement of the latch arm. Attach the latch catch to the box side with two ½" #8 round-head wood screws so the door compresses the weatherstrip equally all around when closed. Don't make the compression excessive; all that's needed is full all-around door contact and slight compression for a good seal. Position the top round-head screw to also act as a latch lever stop. Adjust the L-handle position and fasten with the supplied setscrew.

8. ADD SIGNAGE AND GRAPHICS

You can order vinyl lettering and graphics from any sign company, using the photos from this article as a guide. They can apply the graphics for you, or you can save money by applying them yourself; they'll give you instructions. I masked and spray-painted both box sides with several coats of day-glow orange (you could also use the same red color as the alarm box) and applied the white vinyl "RESCUE" lettering myself.

The 4"×11" graphic of a bystander throwing a Rescue Stick to a swimmer was enlarged by the sign company from an image in the Stick instruction manual. I added a "Sticks are ready to throw as is" message inside on the box back wall because in testing two people asked if the yellow thing on the stick should be unwrapped first.

9. TUCK THE T-HANDLES

Each Stick has a small red T-handle for manual inflation, which we *do not* want used in this application. Tuck the T-handle into the folds of the fabric to hide it (Figure **I**).

LET'S SAVE SOME LIVES

Your Rescue Box can be mounted, centered about head height, on a pier piling or on a 10' 4×4 treated post dug into the ground. Use two 2½" hanger bolts, with wood-screw thread on one end and ¼-20 machine thread on the other. An easy non-measuring way to mount the box is to drive two nails partway into the post 3" apart and level. Rest the box temporarily on the nails while you center it on the post and drill two pilot holes for the hanger bolts through the back of the box into the post. Take down the box and remove the nails. Enlarge the holes

in the box just enough to clear the hanger bolts. Use two nuts jammed together on the end of the bolts to turn the bolts with a wrench and screw them into the pole. Remove the jam nuts, hang the box on the bolts, and secure it with 1½" washers, lock washers, and nuts.

Remove the Mustang Rescue Sticks from their clear protective plastic bags before placing them in the box. Close the box door, remove the stopper alarm pin to arm it, close the clear stopper cover, and the box will stand ready for fast access to save a life.

THE FIRST TEST BOXES

I installed two Rescue Boxes (the first of their kind in the world) on the Bogue Inlet Pier at Emerald Isle, North Carolina, last season for testing. The pier owners, local swimmers, and fire and rescue people welcomed the idea, and the pier will continue to use the boxes, one mounted above on the pier boardwalk (Figure **J**), and one below on a pier piling adjacent to the popular swimming area (Figure **K**).

For increased visibility I added a triangular day-glow orange bicycle flag on a fiberglass wand, cut down and mounted to the box back (Figures K and **L**).

For more advice on placing your rescue boxes in private or public areas, and how to build a dummy stick for throwing practice, visit the project page online at makezine. com/go/bowie-rescue-box.

BOX MAINTENANCE

The box can be taken down for storage during the winter months in the colder latitudes. The 9V alarm battery should be replaced each swimming season, or a 5-year battery could be used.

Directions for Stick maintenance are included with purchase; they are reusable after inflation with a Re-arm Kit (#MA7206) available online. Using a permanent felt-tipped marker, you might print "Property of [you or your group]" and a phone number on the preserver fabric for return after use. ◉

AUTHOR'S NOTE: Thanks to longtime buddy and frequent *Make:* contributor Larry Cotton for his help and the generous use of his garage workshop.

[+] Please share your build, and get more photos and tips, at makezine.com/go/ bowie-rescue-box.

pin, countersunk on the back of the door if necessary. The latch arm is ½"×⅛"×3" aluminum. There's a flat place on the middle of the knob-set pin, so I made an aluminum clamp ½"×⅛"×1¼" and secured it at the flat spot with two #8 machine screws and lock nuts. Cut off the excess pin length and file the cut end smooth. Between the two machine screws, I tapped the pin to receive a ¾"-long 6-32 machine screw through the latch arm and clamp, to keep the mechanism from ever loosening.

There's a thick ¼" ID × 1" OD washer under both the inside latch mechanism and the outside L-handle as bearing surfaces. If necessary, shim these with more washers of different thicknesses to get a nice snug fit.

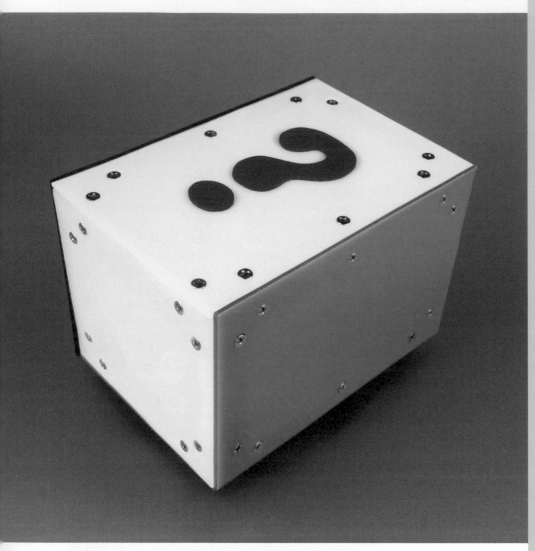

TIME REQUIRED:
A Weekend

DIFFICULTY:
Intermediate

COST:
$35–$60

MATERIALS

» **¼" plywood or ⅛" acrylic** Cut 4 pieces 3"×6" and 2 pieces 3"×3", then trim them to fit together. You'll screw or bolt these into square dowels or plastic angles along the interior edges.

» **¾" square dowels or 1" plastic angles, 4' total length**

» **Screws or bolts, ½" #6 flat-head (60)**

» **DC motor, 6V–12V, approx. 1" diameter, 250–1,000 rpm**

» **Pulley or hub** to fit your motor, from a source such as RobotShop, robotshop.com

» **Bolt, ⅝" #2, with 2 washers and 2 nuts secured with LocTite** so that they don't work loose

» **Galvanized wire, heavy gauge, 1" length**

» **Piezo beeper or buzzer, 6VDC–12VDC** If the specification lists an audio frequency, the beeper generates its own sound, which is what you want.

» **Loudspeaker, 2"–3" diameter, 8Ω impedance or higher**

» **Mercury switches (6)** Cheap on eBay and hard to find elsewhere, as they are environmentally incorrect (don't throw them in the trash!). Or substitute a tilt sensor such as Adafruit #173; it will cost more, and I don't think it's quite as reliable, but you have the option.

» **Scrap wood and sheet metal, small pieces** A steel food can lid will work fine.

» **Texas Instruments 555 timer chips (5)** The 7555 is *not* recommended, as its output may be insufficiently robust.

» **Relays, DPDT or SPDT, nonlatching (3)** coil rated for 4VDC or 4.5VDC, to switch at least 500mA

» **Diodes, 1N4001 (6)**

» **Resistors: 47Ω (1), 10kΩ (5), 68kΩ (1), 100kΩ (1), 150kΩ (1), 330kΩ (2), 680kΩ (1), and 1MΩ (1)**

» **Capacitors: 0.01µF (1), 1µF (3), 10µF (5), and 100µF (1)**

» **Battery carrier for 4 alkaline C cells** or optional 12V lithium battery pack and 2-amp fast-blowing fuse with holder

» **Vibration sensor** I used the "slow" variant, Adafruit #1767; the "medium" and "fast" types are easier to trigger.

» **Throttle spring** from any auto parts store

» **Perf board** to solder a vibration-resistant version of the circuit

» **Hookup wire**

TOOLS

» **Solderless breadboard** for assembly and testing
» **Soldering iron and solder**
» **Wire cutters and strippers**
» **Multimeter**
» **Pliers**

Jumping Mystery Box

Use simple sensors to detect 3D motion and trigger movements and sounds, no processor necessary **Written and photographed by Charles Platt**

The Mystery Box sits on the table, doing nothing. The fun begins when someone picks it up, turns it over, or shakes it. The box screams in protest and starts vibrating as if something inside it is trying to get out. Put it down and leave it alone, and the box returns to its original inscrutable state — until someone else tries to mess with it.

SOLVING THE MYSTERY

Evidently the box detects motion. You can use an accelerometer for this purpose, but I chose something older, cheaper, and simpler: a mercury switch. This consists of a blob of mercury in a glass capsule, as in Figure Ⓐ. When the capsule is tilted, the mercury rolls down and makes a connection between two interior contacts. If you glue the capsule to the inside of the Mystery Box, you have a simple tilt sensor.

Figure **B** shows how the switch can trigger some action for a fixed interval. The switch applies power to a relay, and the relay powers a 555 timer while also running a motor that creates vibration. The trigger pin of the timer is connected to a resistor-capacitor combination (an RC network) that provides a brief low pulse when power is switched on. (To learn more about timers and RC networks, you might like to read my book *Make: Electronics*.)

The output from the 555 connects back to the relay coil, so that the timer keeps the relay contacts closed even when the mercury switch reverts to its "off" status. At the end of the cycle, the timer stops powering the relay, the relay switches off the timer, and the circuit goes back to consuming no power at all. You can leave the box lying around for days or months without it running down its batteries.

The two diodes in the schematic prevent the 6VDC input from fighting with the output of the timer, which has a lower voltage. The diodes themselves impose a voltage drop, so I suggest using a 4.5VDC or 4VDC relay.

REFINEMENTS

The box will start making noise every time someone picks it up and turns it 90° or more along any horizontal axis. I wanted it to behave less predictably, so I combined three mercury switches by gluing them into the faces of a 1" cube, as in Figure **C**. Then I wired them in series so that all of them must close to start the timer.

This still seemed a bit too predictable, so I decided that the circuit should pause for a couple of seconds before making noise, and should then become unresponsive for a couple more seconds at the end of its cycle. Figure **D** shows how three timers and relays can achieve this.

Figure **E** shows the actual schematic. The extra diodes across the relay coils and the motor are to prevent back-EMF that may disturb the timers. I also added a 10µF capacitor between the power supply of the first timer and ground, to protect it from voltage spikes caused by the switches.

To make the Mystery Box shake or jump, I used a DC motor that vibrates because it has a nylon pulley with a bolt inserted off-center. Then I added a pivoting length

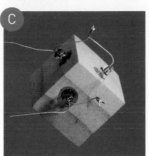

A blob of liquid mercury makes a connection between two contacts (left) or leaves them exposed (right) depending on the angle of the glass capsule.

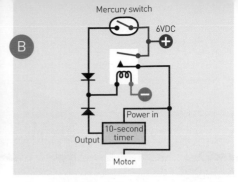

Concept for a circuit that uses no power until the switch closes, then runs a motor for a fixed period, then switches itself off.

Three mercury switches, wired in series, embedded in a 1" cube at 90° angles to each other.

CHARLES PLATT is the author of *Make: Electronics*, an introductory guide for all ages, its sequel *Make: More Electronics*, and the 3-volume *Encyclopedia of Electronic Components*. His new book, *Make: Tools*, is available now. makershed.com/platt

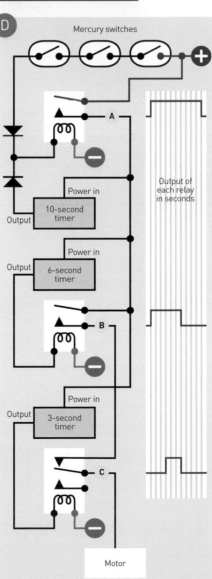

Concept for a circuit that pauses before and after running a motor.

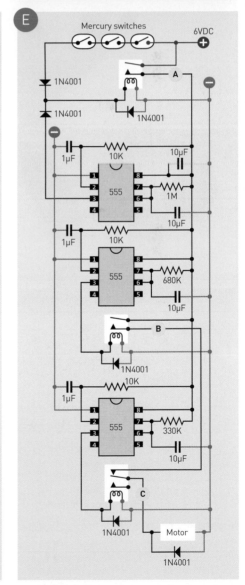

Schematic for the circuit suggested in Figure D.

The DC motor is clamped between two ¾" square dowels.

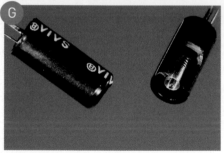

A vibration sensor (left) is cut open (right) to show the tiny rod and spring inside.

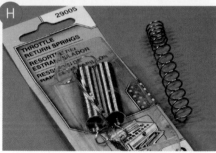

If a vibration sensor is inserted in a throttle return spring, suitably stretched, it will become more sensitive.

of heavy-gauge galvanized wire that bangs into a piece of stainless steel to create a hammering sound (Figure F).

I also wanted the box to protest if someone simply jolts it, so I added a vibration sensor and a beeper. The sensor looks like a small capacitor, but if you slice it open as in Figure G, you find a pin in the center of a coil of wire. When the coil vibrates, it touches the pin. To increase its sensitivity I mounted it inside a throttle return spring that I found in an auto parts store. I stretched the spring so that it would flex more freely (Figure H).

The output of a vibration sensor is very brief, so it triggers another timer which sends a 1-second pulse to a beeper. The additional schematic for this module is in Figure I. By driving it from point A in the main schematic, I bypassed the pause so that the box will protest if it is jolted whenever the electronics are active.

Lastly I wanted the box to play some random musical notes if it is tilted by varying amounts, so I added one more timer, running asynchronously at an audio frequency determined by three more mercury switches. The schematic for this module is in Figure J. It is powered from point B in the main schematic, so that it will start out of sync with the motor.

Because the motor draws significant power when it vibrates, AA batteries were inadequate. I used four C cells in a plastic carrier (Figure K).

Figure L shows the box under construction. I used screws and bolts to hold it together, because glue tends to come unglued if an object is likely to be banged or

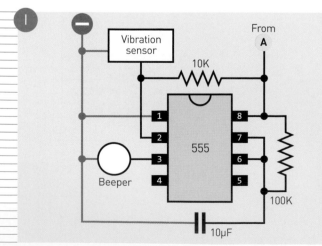

Circuit module to sound a beeper for about 1 second when triggered by a vibration sensor.

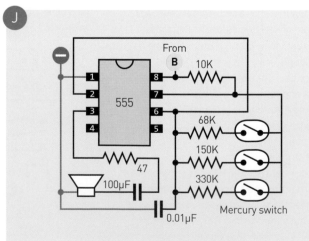

Circuit module to create a variety of audio tones using three mercury switches.

Battery carrier with four C cells.

shaken vigorously.

If you're willing to take more time, a wooden box with an antique look might increase the air of mystery, like something out of an Edgar Allan Poe story. I leave that to you.

OTHER IDEAS

Sensors are cheap, especially from hobby sources such as Electronic Goldmine, All Electronics, SparkFun, or Adafruit. To make the box respond in additional ways you could add an electret microphone, a phototransistor, or perhaps a rotational encoder that is moved by a sinker weight of the type sold for fishing (Figure M).

Other sensors on breakout boards are shown in Figure N. Each of them cost only $4 when I found them online.

If you're willing to spend an extra $20 you can substitute a 12V lithium battery pack for the C cells. This can be recharged through an external socket, so that the box never has to be opened. The extra power will allow you to use a bigger motor to make your mystery box really jump — along with anyone who ventures to touch it. You'll need to substitute 12VDC relays, but the other components can remain unchanged. Since lithium batteries can start a fire if they're overloaded, please include a 2-amp fast-blowing fuse. ●

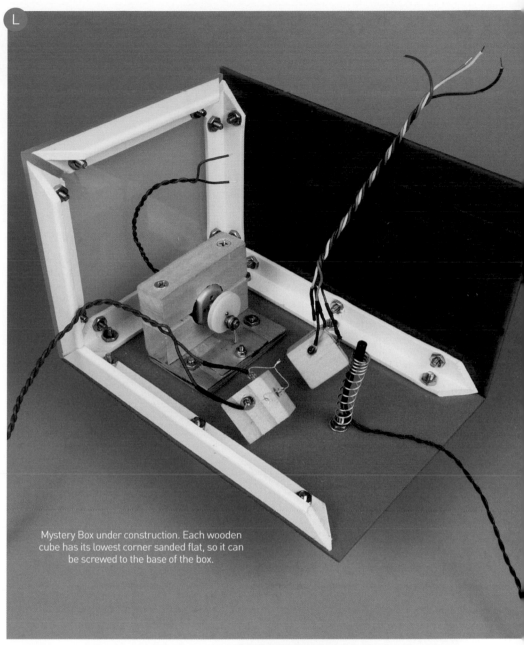

Mystery Box under construction. Each wooden cube has its lowest corner sanded flat, so it can be screwed to the base of the box.

A fishing sinker weight will turn a rotational encoder if the encoder is shaken or inverted.

Low-cost sensor breakout boards — tilt sensor (left), bump sensor (center), and touch sensor (right).

Ionizing Radiation

Measure the background radiation in your environment with a Geiger counter

Written by Forrest M. Mims III

FORREST M. MIMS III an amateur scientist and Rolex Award winner, was named by *Discover* magazine as one of the "50 Best Brains in Science." His books have sold more than 7 million copies. forrestmims.org

TIME REQUIRED:
0–16 Hours

DIFFICULTY:
Easy

COST:
$100 and up

TOOLS
» **Geiger counter** You can make one from scratch (makezine.com/projects/geiger-counter), build a kit from imagesco.com or mightyohm.com, or buy one ready-made.
» **Computer with internet connection** for sharing your radiation data

We and everything around us are continuously exposed to a wide range of electromagnetic radiation and subatomic particles, including radio and microwaves, byproducts of cosmic rays from outer space, and radioactive emissions from rocks, soil, building materials, and some foods. Even fireplace ash and the potassium in our bodies are slightly radioactive.

TYPES OF RADIATION

Electromagnetic radiation ranges from very long radio waves to very short x-rays. In between are ultraviolet radiation, visible light, infrared, and microwaves.

Very short wavelength UV radiation, x-rays, and fast-moving, subatomic alpha and beta particles have enough energy to dislodge electrons from atoms. The resulting atom is called an *ion*, and the radiation that forms ions is called *ionizing radiation*.

Cosmic rays are very high energy particles from outside the solar system that create showers of subatomic particles when they strike molecules of air in the upper atmosphere. These ionizing particles are called *muons*. While muons, like electrons, have a negative charge, they have far more mass than electrons.

Soil and minerals that emit ionizing radiation are described as *radioactive*. Uranium is the best-known radioactive mineral.

Because ionizing radiation can damage DNA and lead to cancer and other health issues, there is a general concern that all radioactive exposure is harmful. Those who are exposed to high levels of radioactivity have good reason to be concerned. The rest of us have little to fear from normal and unavoidable exposure to slight amounts of radiation plus occasional dental and medical x-rays.

DETECTING IONIZING RADIATION

Several methods are used to detect ionizing radiation. The oldest detector is the *Geiger–Müller (G-M) tube*, which was developed by Hans Geiger and Walther Müller in 1928. A G-M detector is a hollow metal tube with a central wire electrode. The tube is filled with a gas such as neon or helium under low pressure, and several hundred volts are applied across the tube and the electrode. When an ionizing particle or ray enters the tube, the gas inside is briefly ionized, and this provides a brief, electrically conductive path for the voltage across the tube. This voltage pulse can be counted and amplified to provide an audible clicking sound from a speaker. G-M tubes can detect beta particles, muons, gamma rays, and x-rays. By installing a mica window in the end of the tube, alpha particles, which can be blocked by paper, can also be detected.

Some crystals and plastics known as *scintillators* emit a flash of light when struck by ionizing radiation. These materials are installed inside a light-tight enclosure along with a photodiode to provide a sensor for detecting beta particles, x-rays, and gamma rays. Various *semiconductor diodes* can also detect ionizing radiation, but the old-fashioned Geiger tubes are much more reliable. That's why most commercial Geiger counters use them.

SELECTING A GEIGER COUNTER

Many kinds of Geiger counters are available at prices ranging from less than $100 to more than $1,000. John Iovine described a DIY Geiger counter in *Make:* Volume 29 (makezine.com/projects/geiger-counter). His company, Images Scientific Instruments (imagesco.com), sells a wide range of assembled and kit Geiger counters. Many other kits and assembled counters can be found online; for example, the MightyOhm Geiger Counter kit (mightyohm.com/blog/products/geiger-counter) is about $100.

The *counts per minute (cpm)* measured by a Geiger counter can be converted into cumulative radiation exposure, which is known as the dose. The traditional dose unit is the *roentgen (R)*, and the International System of Units (SI) unit is the *sievert (Sv)*. The manual for the popular RadAlert Geiger counter gives the conversion of cpm to these units when exposed to gamma rays from radioactive cesium 137 (geigercounters. com/content/Rad100XManual.pdf). The following table was extrapolated from this manual (µSv is microsieverts and mR is milliroentgens):

CPM	µSv/hr	mR/hr
12	0.1	0.01
120	1	0.1
1,200	10	1

Specifying radiation intensities and doses is a complex topic covered in detail by many websites.

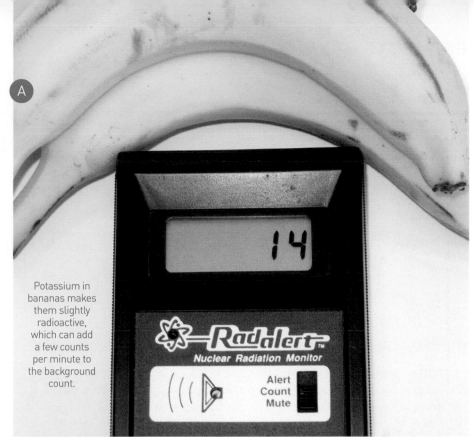

Potassium in bananas makes them slightly radioactive, which can add a few counts per minute to the background count.

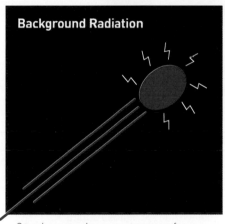

Cosmic rays and gamma rays come from rocks, soil, and outer space.

Radioactive radon gas is emitted by the hot springs at New Mexico's Soda Dam.

RADIOACTIVE MATERIALS

The radium in Brazil nuts and the potassium in bananas make these foods slightly radioactive (Figure Ⓐ). The orange color of old china saucers and plates was often produced by radioactive uranium oxide. The emissions from coal-fired power plants are slightly radioactive. Granite countertops, fireplace ashes, the fine clays used to coat the pages of glossy magazines, and the smoke detectors that warn about fire are all very slightly radioactive.

Radioactive radon gas in soil can collect in basements of buildings. The air over the natural hot springs at New Mexico's Soda Dam is slightly

Geiger counter background count network operated by GQ Electronics (© 2018).

radioactive (Figure B previous page). Accidental radiation can be released by nuclear power plants and dispersal of radioactive soil from uncovered trucks. Nuclear power plant disasters, "dirty bombs," and nuclear explosions can spread radioactive contaminants over a wide area.

GEIGER COUNTER EXPERIMENTS AND PROJECTS

The most basic application for a Geiger counter is measuring the background radiation count at a location. The background count detected by a standard G-M tube is typically 10–20 cpm. The background count for a G-M instrument with a larger "pancake" detector is typically 30–60 cpm.

You can share your Geiger counter data with networks of background counters across the United States and other countries. These networks will become very important should there be a nuclear power accident or other major source of unwanted radiation. Two such networks are at radiationnetwork.com and gmcmap.com (Figure C).

The average background count measured by a RadAlert Geiger counter at my Texas site is 11 cpm. The background count in Fort Stockton, Texas, is around 18 cpm and that in Roswell, New Mexico, is around 22 cpm. The count is higher than these numbers halfway between my site and the Texas coast because of uranium occurring naturally in the soil.

A Geiger counter recorded background count samples when I drove from Central Texas to Los Angeles and back to study the solar eclipse of July 11, 1991, from a cruise ship. The average background count across the southwestern United States was 16.5 cpm. At sea the average was only 9 cpm (Figure D).

Muons are absorbed by the atmosphere, so there are many more of them high in the sky than at the surface. This quickly becomes obvious if you take a Geiger counter on a drive through mountainous terrain. While flying in commercial jet airliners, I've measured cosmic ray intensities 30 to 40 times higher than on the ground. Figure E shows the significant changes in background count as a plane changed altitude during flights from San Antonio to Atlanta and then to Zurich, Switzerland.

If you try this, be sure to mute the Geiger counter's speaker. That's because the count will be so rapid at high altitudes that a Geiger counter will emit a buzz. As I've learned more than once, this can alarm both fellow passengers and flight attendants.

GOING FURTHER

Exposure to low levels of radioactivity is a part of everyday life. During the time you've spent reading this article, many thousands of ionizing rays and subatomic particles raced through your body. Measuring them and understanding their health effects is a complex subject, and you can learn much more in a careful online search. ◐

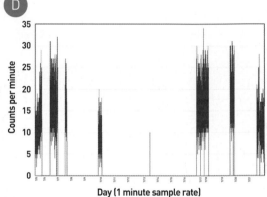

Background count measured by a RadAlert Geiger counter from Texas to a cruise ship off Baja California, and back.

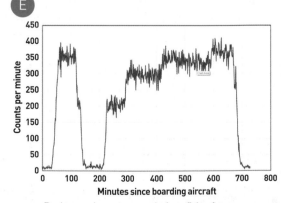

Background count on an airplane flying from San Antonio to Atlanta and a second aircraft from Atlanta to Zurich.

CAUTION: *Make:* and I make no warranties (express or implied) about the information provided in this article. Always use care when monitoring radioactivity that exceeds the usual background count. Should you encounter a persistently high background count, move away from the area and notify a health or environmental agency — as I once did when responding to a drug store's request to check out a highly radioactive source once sold as a health remedy.

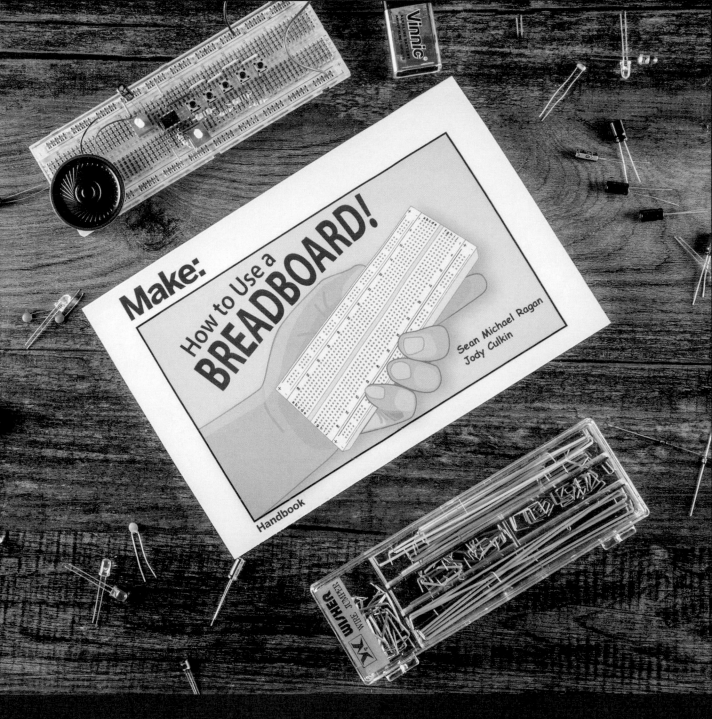

Sun-Fueled Fun

Written and
photographed by
Andy Forest

Solar panels + bilge pump = nonstop running water for boats, castles, and water fights!

The Solar Beach Hose is a solar panel–powered bilge pump connected to a garden hose. It's the absolute favorite beach toy for my kids, their friends, and their cousins. They make rivers to float boats down, fill buckets with water for making sand castles, spray each other, and have lots of fun. Ever since I made it, everyone that walks by on the beach wants one!

The solar panels are the type that are widely available for boats, RVs, and camping, and the whole project is very easy to make using inexpensive parts from the hardware store — no soldering is required.

MAKE YOUR SOLAR BEACH HOSE

We're working with low voltage with this project, so electrical safety is not a concern. The 12V solar panels put out enough voltage to give you a little tingle, but are perfectly safe. We are using some components from 120V house wiring like the light switch, but that's just for a reliable and convenient enclosure.

1. PREPARE THE CABLES

Your first step is to hook up the solar panels to the light switch. I kept the quick-connect adapters intact that come with the Coleman panels so I could hook them up to the included battery charge controller.

Cut the large alligator clamps off the quick-connect cables (Figure Ⓐ), and strip about ½" of insulation from the cut ends of the wires. (Other panels may already have a cable that terminates in stripped wires.)

NOTE: On my first build, I cut the solar panel quick-connect extension cord in half to use the connectors on both ends for this step. Don't do it! This gave me two identical-looking connectors with reversed red-black wiring polarities. It's really confusing to keep positive and negative consistent in this situation.

TIME REQUIRED:
1–2 Hours

DIFFICULTY:
Easy

COST:
$150–$165

MATERIALS
You can get most of this at your local hardware store. I bought all but the bilge pump from Canadian Tire!

» **Folding solar panels, 35W–40W, 12V (2)** I used Coleman/Sunforce #38040, Canadian Tire #2995775, canadiantire.ca. But this model's getting hard to find; Amazon #B01G1II6LY looks like a good substitute.
» **Bilge pump, 12V** such as Amazon #B00N6HDRLS
» **Lamp cord, 18 gauge, 2 conductor, 25' long** Canadian Tire (CT) #0529969
» **Wire connectors (4)** aka wire nuts, CT #0522092
» **Electrical utility box, 1-gang, metal** CT #0521210
» **Light switch, standard** CT #0528204
» **Switchplate, 1-gang, metal** CT #0526747
» **Utility box connector** CT #0521827
» **Garden hose, 15' long** CT #0591104
» **Faucet hose adapter** to connect bilge pump to hose, CT #0633211
» **Tent peg** CT #0765745

TOOLS
» **Wire strippers**
» **Electrical tape**
» **Screwdriver**
» **Scissors**

ANDY FOREST is co-founder and executive director of steamlabs.ca, developing kids' identities as creators, coders, scientists, and makers in the 21st century. He consults with science centers and educators around the world including Toronto, New York, San Francisco, Oslo, and London.

To prepare the lamp cord, use the scissors to separate the two conductors and then strip ½" of insulation from both wires. Repeat at the other end.

2. PREPARE THE ELECTRICAL BOX
You're going to hook up all the wires inside the metal electrical utility box. First, remove one of the metal circle "knock-outs" by pushing it in with a screwdriver and then pushing it back and forth until it breaks off.

The utility box connector will protect the wires from being pulled out, or cut by the sharp edges of the box. Unscrew the nut, fit the connector into the box, and attach the nut from the inside (Figure B).

Feed the solar panel connector wires and lamp cord wires through the hole before connecting them since they won't fit through afterward (Figure C).

3. CONNECT WIRES TO SWITCH
One wire of the lamp cord is ribbed, and one is smooth. Use the ribbed side as negative to keep the polarity straight. Twist together the 2 black wires from the solar connectors and the ribbed cable of the lamp cord, and secure them with a wire connector, aka wire nut. Push the wire nut firmly onto the twisted wires while turning it clockwise; there's a copper spring inside that grabs onto the wires and keeps them secure. Then tug each of the 3 wires individually to make sure they are firmly in place and won't wiggle loose.

Cut a small jumper wire from the lamp cord, then twist it together with the two red wires from the solar panels, and secure with a wire nut.

Make little hooks out of the smooth, positive lamp wire and the jumper to connect them to the light switch terminals. Bend the hooks clockwise, so that tightening the terminal screws doesn't unravel the wires (Figures D and E).

4. CLOSE UP THE BOX
Pull the slack in the wires out of the box, and tighten the screw in the utility box connector to clamp them in place (Figure F). Screw in the light switch and then the faceplate (Figure G).

5. CONNECT THE BILGE PUMP
Now you'll hook up the bilge pump to the other end of the lamp cord. Use a wire nut

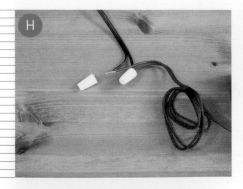

to connect the ribbed lamp wire to the blue or black negative wire. Connect the smooth lamp wire to the pump's red positive wire (Figure H).

Since the bilge pump and its wires will be in the water, protect the wire connectors by wrapping them in electrical tape (Figures I and J).

6. CONNECT THE HOSE

Depending on the size of the outlet on your bilge pump, you may need to connect the faucet hose adapter on the inside or the outside of the outlet. I had to take off the metal ring clamp to make it fit inside the outlet of my pump (Figure K).

Then the garden hose can screw into the adapter and you're ready to go !

GET PUMPED!

Place your solar panels facing directly into the sun, near your towels on the dry part of the beach. Turn the switch off and connect the quick-connect adapters.

Put the bilge pump in the ocean or lake, and anchor it in place with the tent peg (Figure L). These pumps are meant to take a lot of abuse, and pumping sand-filled seawater doesn't seem to faze them.

Bury the lamp cord in the sand, since it could be a tripping hazard for people walking along the beach.

As the tide goes out, the bilge pump will even excavate a little hole for itself, and happily keep on pumping. If your beach is really gently sloped, you can dig a little hole for the pump in the sand to give it access to more water to pump (Figure M).

Kids will spend hours with the hose, using it for water for sandcastles, making rivers, excavating, melting sand. It opens up so many new ways to play! Slip and slide at the beach, anyone (Figure N)?

MAINTENANCE

Before you leave, let the pump run in clear water for a bit to clear out any sand. When you get home, if you were pumping seawater, rinse out the pump with fresh water to get rid of the salt to prevent corrosion.

GOING FURTHER

Any cloud cover at all will make the output of the solar panels drop to nothing and the pump will stop. You can add in a battery backup to prevent this! These solar panels come with a charge controller, so you can just buy a waterproof sealed marine battery and hook it up. ✎

[+] See more step-by-step photos at makezine.com/go/solar-beach-hose.

1+2+3 Glass Block Night Light

Written and photographed by Juliann Brown

There's something very relaxing about fairy lights, but only Carrie Fisher could get away with keeping her Christmas tree up all year. This project lets you keep the festive feeling alive year-round — plus it's a great night light or custom gift.

1. STUFF IT

Clean off your glass block to ensure a pristine surface for painting. Cut off sprigs of faux greenery (Figure A) and stuff them into the block's opening as you feed in the fairy lights (Figure B).

> **TIP:** Don't fill the block too tightly or the lights won't show through.

2. PAINT IT

Tape your stencil onto the front in the desired position and paint your silhouette (Figure C), then let it dry. Optionally, you could purchase or make a vinyl silhouette sticker.

3. BLING IT

Pour in about one teaspoon of glitter (Figure D) and carefully move it around the inside of the block to coat the glass surfaces. Keep the block opening upward so you don't end up finding glitter between your toes later.

Crafting glass blocks usually come with a removable plastic stopper that you can thread the cord through with the help of a scissor cut (Figure E). Seal the hole in the stopper with some tape and turn on the fairy lights (Figure F)!

GOING FURTHER

Get even more creative with your interior scene. My next one will have a prehistoric setting full of little dinosaurs, or a cool jungle backdrop with a handful of monkeys. ⊘

TIME REQUIRED:
1 Hour

DIFFICULTY:
Easy

COST:
$25–$50

YOU WILL NEED

- » **KraftyBlok** or similar crafting glass block with a bottom cutout and removable seal
- » **Faux greenery**
- » **LED string fairy lights**
- » **Glitter**
- » **Black acrylic paint**
- » **Paintbrush**
- » **Stencils** for your silhouette
- » **Wire cutters**
- » **Scissors**
- » **Scotch tape (optional)** but highly recommended to help seal off the bottom to avoid annoying traces of escaped glitter

JULIANN BROWN is the art director for *Make:* magazine.

A

B

C

D

E

F

Scrappy Circuits

Hack dollar-store tea lights to make affordable electronics modules for learning

Written by Michael Carroll

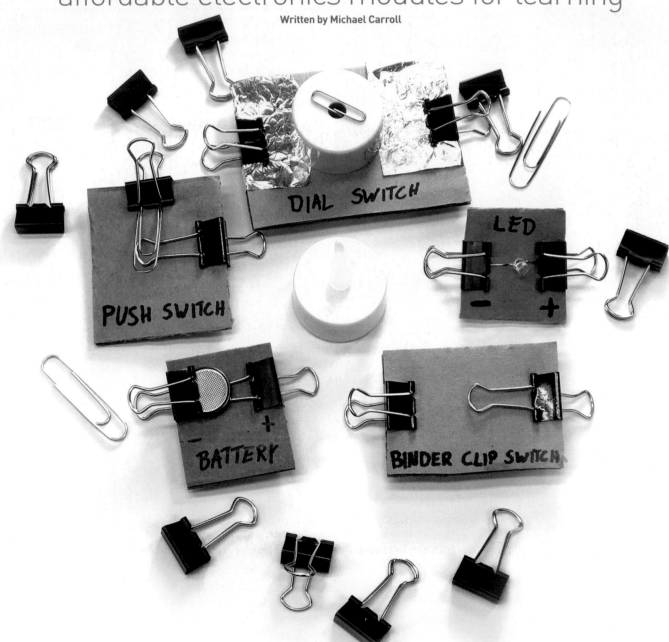

DIAL SWITCH

LED

PUSH SWITCH

BATTERY

BINDER CLIP SWITCH

The best dollar you can spend on a child's STEAM education is to take them to the dollar store, buy an electronic item, and then take it apart. Learning how a handheld fan, LED light, remote control, or headphones work will help show how wind turbines, traffic lights, speakers, keyboards, and other modern devices function.

The next step is to take apart a dollar store item and make it better — combine two items, replace a switch, create something new and unique. Scrappy Circuits is the perfect way to get started.

I first developed Scrappy Circuits with Chris Connors and Eva Luna while working at Xraise, an education outreach program

at Cornell University. What began as an exploration of DIY switches ended later as a self-made invention system sourced from a fifty-cent LED tea light. Educators can build it with their class for about $20. Anyone with access to office supplies and a dollar store can make the five Core Bricks for about $1. Here's how to create each one.

TAKE APART A TEA LIGHT

Open the battery compartment, remove the 3V coin cell battery, and pry the base/LED holder apart from the outer cylinder body (Figure Ⓐ). You'll use three parts: the LED, battery, and cylinder body. For a simple demonstration of how electricity travels, make the LED legs straddle the battery, with the long leg touching the positive (+) side.

Battery Brick

The *power* source for your Scrappy Circuits.
1. Clip a binder clip to one side of a rectangle of cardboard (a "brick").
2. Flip one binder clip arm down.
3. Trim the opposite side of the cardboard slightly longer than the extended arm.
4. Place the battery on top of the clip arm with the smooth, positive (+) side down.
5. Clip a second binder clip over the battery.
6. Label each clip positive (+) or negative (–) based on which side of the battery it touches. This makes lighting the LED easier later.

LED Brick

Of the Core Bricks, this is the only *load*, or item that consumes electricity.
1. Stretch the legs of the LED apart.
2. Clip a binder clip around the longer LED leg and the cardboard. Label this side (+).
3. Clip the opposite leg and label it (–).

Binder Clip Switch

It operates just like a *toggle* light switch.
1. Clip two binder clips to opposite sides of a brick.
2. Flip one arm flat against the brick. It should not touch the other binder clip. If it does, cut a larger brick.
3. When you flip the opposite arm down, it completes the circuit (turns on). When you lift the arm, it turns the circuit off.

Push Switch

Momentary switches turn on when pushed down. When released, they turn off again.
1. Clip one binder clip to a cardboard brick.
2. Lower the upper arm.
3. Lay a paper clip perpendicular over the binder clip arm.
4. Use a second binder clip to hold the paper clip in place.
5. After it is clipped, bend the paper clip up slightly so it's *not* touching the other arm.

6. Push the paper clip down to close the switch and turn your LED on.

Dial Switch

A creative use of the tea light cylinder body.
1. Use glue stick to attach aluminum foil to each side of a cardboard brick. Be sure the two pieces do not touch in the center.
2. Add binder clips to each end.
3. Glue foil to 50%–75% of the inside and outside of the lower edge of the cylinder.
4. Poke a hole through the cardboard.
5. Insert a straightened paper clip through the cylinder and the hole.
6. Bend the paper clip to capture the cylinder. Tape it down or cut off excess.
7. Spin to turn on or off. When the foil on the cylinder touches both pieces on the cardboard, the circuit will turn on. When it touches just one side, it will be off.

Scrappy Clips

Your bricks can be connected with alligator clips or with Scrappy Clips. To make a Scrappy Clip, cut 6"–10" of aluminum foil. Fold it in half a few times and then roll it tight. Hook each end through the one-hoop end of a small paper clip. Squeeze the connection tight with pliers (Figure Ⓑ).

MAKE A CIRCUIT!

Gather two Scrappy Clips, the Battery Brick, and the LED Brick. Slide a Scrappy Clip onto the binder clip to connect your bricks (Figure Ⓒ). Connect the positive (+) side of the battery brick to the (+) side of your LED brick. Next connect the two remaining sides to illuminate your LED. (If it doesn't light, try sanding the mouth of your binder clips and squeezing each connection.) Take a moment to appreciate how cool this is. You just created an electrical circuit!

Now get one of your switch bricks to learn how you can control your circuit. Disconnect any clip and reconnect it to your switch. Use another Scrappy Clip to connect the remaining sides (Figure Ⓓ). The switch now controls the LED light.

WHAT'S NEXT?

Make a Buzzer, Magic Wand, Light Sensor, and more at www.scrappycircuits.com. And please consider supporting our Kickstarter campaign to bring Scrappy Circuits to more inventors-in-training around the world. ◗

TIME REQUIRED:
1–2 Hours

DIFFICULTY:
Easy

COST:
$1

MATERIALS
» **LED tea light** Two for a dollar at any dollar store.
» **Small binder clips (10)**
» **Paper clips (2)**
» **Cardboard**
» **Aluminum foil**

TOOLS
» **Scissors**
» **Pliers**
» **Glue stick**
» **Small screwdriver** to pry open tea lights
» **Pen or marker**

MICHAEL CARROLL is a writer and third grade teacher at Overlook Elementary School in Abington, Pennsylvania. He has a passion for high-interest, low-cost projects so that *all* students can learn.

Colleen Coyne, Michael Carroll

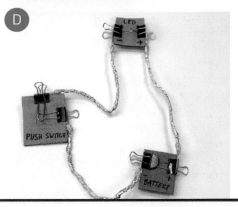

[+] For more photos and tips, visit the project page at makezine.com/go/scrappy-circuits.

Garden Gizmos

AVOCADO BOAT
YOUR TOAST WILL THANK YOU.

makershare.com/projects/avocado-boat

This is a little boat designed to carry an avocado pit to help it germinate by keeping its bottom immersed in water but its top dry. Before, I tried to germinate them using wooden skewers on top of a glass, but the problem was the water level — it evaporated fast, especially in summer, and it takes 4 to 8 weeks for the pit to split open and start growing roots. You can leave this boat floating on water so you never have to worry about the water level.

I designed the boat in Rhinoceros, and 3D printed several iterations to make it float properly. I shared the final model on Thingiverse and it has been downloaded over 8,000 times. I ended up adding my makerspace watermark on the back of the boat because people were selling them on eBay, Etsy, and other websites! —*Igor Daemen, Eindhoven, The Netherlands*

3D-print these gadgets to upgrade your green space

DIGITAL SUNDIAL
DOT-MATRIX AMAZEMENT.
makezine.com/go/mojoptix-sundial

Last issue we told you about the high-tech sundial designed by engineer "Julldozer" in Toulouse, France — it's 3D-printed with hundreds of precisely angled holes that project tiny dots of sunlight to display the current time in numerals, every 20 minutes. Astounding!

I looked for remixes on Thingiverse and found an upgrade that makes it look even nicer in your garden: a gracefully curved, two-legged base to replace the weighted jar. My favorite is Pavel Janacek's version, thingiverse.com/thing:2402017. —*Keith Hammond*

CUTE CLASSICAL BAROMETER
KEEP A WEATHER EYE.
makershare.com/projects/3d-printed-cute-barometer

I had this idea of creating a design of a classical membrane barometer that I could print. So I did it!

The design consists of a 3D-printed flange that squeezes a rubber membrane between its two halves. On one side of the membrane is a sealed jar that's used as a constant pressure chamber; on the other side is the atmospheric air. Any pressure difference between them results in membrane movement, which determines the relative position of a needle on a scale (from sunny to stormy weather).

It took a while to make sure that the pressure flange is actually sealed to the jar so the constant pressure chamber does not communicate to the atmospheric pressure, but once I managed it, it worked like a charm. In order to achieve good airtightness, I recommend printing the layers with 0.2mm resolution. —*Marius Taciuc, Iasi, Romania*

FENCE POST HATS
DEFEAT DRY ROT, IN STYLE.
makershare.com/projects/fence-post-hat

A simple hat for a fence post to guard against moisture — with some stylish roof curves. It's built for a wooden post 75mm square, but feel free to scale it up and down to adapt it to your post size (e.g., 89mm for a nominal 4×4 post).

Normally I would prefer printing in PLA because it can be recycled, but I recommend you use PETG for this project, as it is more durable and robust for outside usage than PLA. —*Xo Ne Un, Planet Earth*

Igor Daemen, Pavel Janacek, Marius Taciuc, Xo Ne Un

Reproducible Ring
How I designed a custom 3D printed robot claw engagement ring

Written and photographed by Sam Freeman

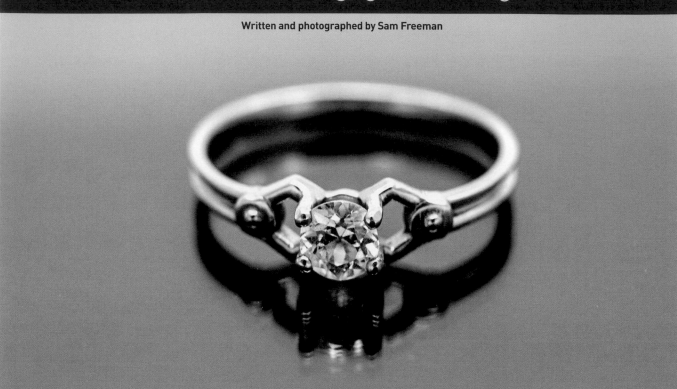

The other day I proposed to my girlfriend with a Ring Pop. She got a kick out of it, it was sweet, and now we have a story to tell. But it was pretty clear that at some point she was going to need a ring that wouldn't dissolve overnight in a glass of water.

She'd been around me long enough to love the idea of wearing a 3D printed ring. If you want major bonus points with your significant other, 3D print a ring that's one-of-a-kind and will last a lifetime, but also is technically reproducible if it accidentally flies out a window or something.

SKETCHING A ROBOT RING

The goal of sketching isn't to get everything perfect, it's to learn as much as possible on paper so there are no surprises during modeling (Figure A). I knew we were going to print the ring, so I thought about how I would build it in software as I was sketching. That was one reason we went with a robot

claw motif instead of an organic form that would've made more sense to create by hand. It didn't hurt that she's memorized every line of dialog in *Futurama*.

MODELING AND TESTING

If the person you're designing a ring for has one that they wear already, you can try to measure it without them knowing and sneakily get their ring size (Figure B). Unfortunately my girlfriend has a variety of rings in various sizes, so I ended up measuring two of them that sort of fit and aimed somewhere in the middle. While I was at it, I measured the width of the rings she liked most to keep all the proportions in check.

I turned to Fusion 360 for a model I could tweak until I got it perfect (Figure C). The program is pretty flexible about allowing you to revise dimensions from earlier. I also like how it takes care of version tracking so you don't have to. It does require an account

to use the free version, but that's easier to swallow than the cost of some software with similar functionality. This was my second project in Fusion 360, so don't let fear of a steep learning curve discourage you.

Depending on your design, you can build a ring out of a single Revolve and a few Extrudes. I could spend this whole article on a Fusion 360 tutorial, but thankfully *Make:* has some wisdom (see *Make: Fusion 360 for Makers*, makershed.com/products/fusion-360-for-makers), and Autodesk has resources (autodesk.com/products/fusion-360/get-started) for learning their program (and those will probably stay up to date). If this is your first time in CAD, you might want to play around with various functions like Mirror and Pattern to see what makes your life easiest.

"Fail early and fail often." When you've got the basic design, it's time to print, deburr, and try it on. I printed out five versions in PLA (Figure D) before sending away for

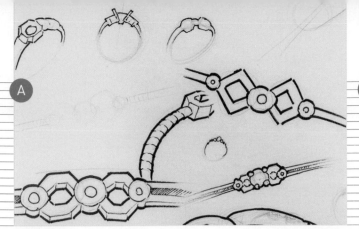

A

B

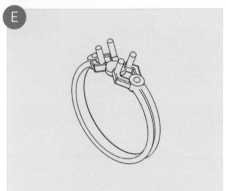

C

D

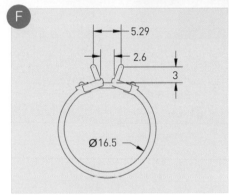

E

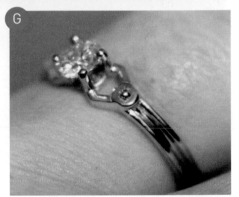

F

G

one in metal. I'm glad I did, because we got useful feedback from people who thought the first version was an infinity symbol. It's easy to get caught up in sub-millimeter details when you're working with a model blown up on a 24" screen, so keeping a 1:1 scale model of your design on hand helps keep your brain grounded in reality.

PRINTING BY MAIL

Third-party printers have access to machines I can only wish to own. I've had great experiences getting 3D prints through Shapeways, so I felt good about using them for this project (not a paid endorsement).

Shapeways spells out design guidelines for each of their materials — maximum size, minimum wall thickness, and so on. You should follow these. Even if you can push the limit and still get a print, the result will be much more durable (and probably better looking) if you stay inside the safe zone. Thankfully their precious metal

products are all produced through lost wax casting, so if your model fits the rules for one metal, like silver, it will (probably) work for another, like 18-karat gold. They don't sell wax models anymore, but they do offer various levels of polish on their prints.

SETTING THE STONE

Setting a stone in a ring is hard (I assume). If you have experience or want to try it for the first time on an engagement ring, go for it.

However difficult it is, I definitely didn't want to screw it up, so I found a local pro to do it instead. He was a huge help. We learned the stone was cut by hand, and he shared design tips, like leaving as much of the stone exposed as possible to let in more light. While he offered to solder a crown on top of our initial silver ring, it would have stuck out and been slightly more cumbersome. The jeweler advised me to spin another ring, this one with built-in prongs and a deeper platform for the stone

(Figure E and F). The final ring is white gold, which makes for more durable prongs than silver (Figure G).

So find someone you trust, and send them drawings of your design before getting an expensive one printed. Fusion 360 has tools that can turn out a dimensioned drawing in 10 minutes. You can expect to pay $150–$200 for the print with shipping (it's okay, she already knows), depending on your design.

Once we had everything in place, the turnaround was quick and the work was excellent. My fiancée is very happy with her ring, and it'll last a lot longer than the Ring Pop did. ✏

marking position for the tyres

Maker Share
Mission to Make:
Editors' Choice

Sound Blink

Written and photographed by Adhith Mathradikkal

A unique DIY portable speaker and LED light show

MUSIC AND LIGHTS HAVE ALWAYS BEEN A FASCINATION FOR ME — they're soulmates in electronics. This project, which I named Sound Blink, is an effort to unite them: a hand-built portable speaker with remarkable sound quality, features that rival premium speakers, and a unique LED light show.

I was adamant on making the build with only simple materials so that even people with no access to CAD techniques could replicate it. This approach challenged my capabilities to find unconventional solutions. The key design aspects are:

» Placement of electronics outside the speaker enclosure, for easy customizations and repair
» Fully screw-fixed assembly and curved wooden enclosure

» Six speaker drivers, a Wi-Fi controlled LED matrix with VU meter, and multiple charging inputs
» User-friendly features like soft buttons for onboard music control, charging level indicators, and group function with AUX output

Planning and figuring out the most effective techniques was more time-consuming than the actual build. The first task was to make the wooden enclosure. It follows a layer enclosure technique in which plywood cutouts are stacked to form a single structure, secured with wood glue and brad nails (Figure Ⓐ).

In order to achieve clear and distinct audio quality, I used multiple speaker drivers (Figure Ⓑ) to enhance low, mid,

and high audio frequencies: passive radiators, woofers, and tweeters, with two-way crossovers. The wooden enclosure gives this speaker a natural and pristine sound quality.

The acrylic casings at the front were formed by warming a couple of acrylic sheets with a heat gun and rolling them onto a custom die made from PVC pipe. The top casing was covered with black sun protection film to resemble a curved LCD screen and provide a light shade to the LED matrix. The bottom one was covered with black jute cloth that conceals all the electronic components. The bottom casing also comes with holes for music control buttons and indicator LEDs.

To make the speaker grille, I drilled hole patterns into a section of PVC pipe, then covered it with speaker grille cloth using upholstery glue. The grille conceals the woofers and tweeters at the rear.

A tapered end cap made from a PVC reducer fitting was placed on the top and the bottom of the final assembly. These protect and cover the passive radiators and also add a nice aesthetic to the build. They're also covered with black jute cloth.

The LED matrix at the front consists of horizontally and vertically placed 12V RGB LED strips. I glued the strips to an OHP sheet (for overhead projector transparencies) so they could easily be fixed around the curved enclosure (Figure C). Strips were wired in a specific pattern for the desired lighting effect and color combinations.

A total of eight different electronic components are used in this project, including a 50W class D amplifier board with Bluetooth 4.0, two buck-boost modules for voltage regulation, a battery monitoring system (BMS) for safe charging and discharging of the battery pack, a Wi-Fi LED driver for controlling the LED matrix, a 4S1P lithium battery pack, and a constant-current, constant-voltage buck converter for battery charging (Figure D).

Since I wanted to combine music with lights in this project, I built a custom VU (volume unit) meter circuit that drives the LEDs to the music. It's based on an LM3915 IC with a dedicated op-amp–based peak detector and output Darlington transistors for better sensitivity and reduced IC heating. I'm pretty happy with the results — it truly added a visual charm to the project.

The acrylic casing, LED matrix, circuit components, and other parts are all screwed to the wooden enclosure. Like the Maker's Bill of Rights says, "Screws better than glues" — this is helpful for future repairs or customization.

Since this project required various processes like woodworking, electronics, and crafting, I was able to expand my knowledge in these areas. Building it was an incredibly rewarding experience. ✏

[+] Get the plywood templates, wiring diagram, and custom VU circuit schematic at makershare.com/projects/sound-blink-unique-diy-portable-speaker-0. And download the Maker's Bill of Rights (aka Owner's Manifesto) at makezine.com/go/makers-rights.

TIME REQUIRED:
2 Months

DIFFICULTY:
Intermediate

COST:
$50–$100

MATERIALS
ENCLOSURE
» **Plywood, 12mm thick** about 120cm× 50cm total area
» **Acrylic sheet, 2mm thick** 31cm×16cm for top, 31cm×9cm for the bottom
» **PVC pipe, 4" diameter** 18cm length for making the die, 23cm for speaker grille
» **PVC reducer fittings, 4" to 3" (2)**
» **Black jute cloth, speaker grille cloth**
» **Sun protection window film, black**
» **Overhead projector (OHP) transparency film**
» **Wood glue, super glue, upholstery glue, double-sided tape, and brad nails**

ELECTRONICS
» **Amplifier board, 25W+25W, with Bluetooth 4.0**
» **Wi-Fi RGB LED driver board**
» **VU meter board** I built this myself; see makershare.com/projects/sound-blink-unique-diy-portable-speaker-0 for instructions.
» **Buck/boost modules (2)**
» **Lithium ion cells, 18650 (4)** The higher the capacity, the better.
» **Buck converter, constant current, constant voltage**
» **RGB LED strips, 12V, 5 meters total length**
» **Two-way audio crossovers (2)**
» **Woofers, 3", 30W, 8Ω (2)**
» **Tweeters, 1½", 30W, 8Ω (2)**
» **Passive radiators, 4" (2)**
» **Hookup wire and switches**
» **Electrical tape and heat-shrink tubing**

TOOLS
» Jigsaw and hacksaw
» Drill with bits, hole saw, sanding wheel
» Belt sander or hand sander
» Heat gun
» Soldering iron
» Rulers, scissors, utility knife
» Glue gun

ADHITH MATHRADIKKAL is a mechanical engineering graduate in India who is fascinated with science and technology, and dreams of working in a mech- or tech-related company. He believes doing DIY projects helps makers understand basic concepts and gain practical knowledge.

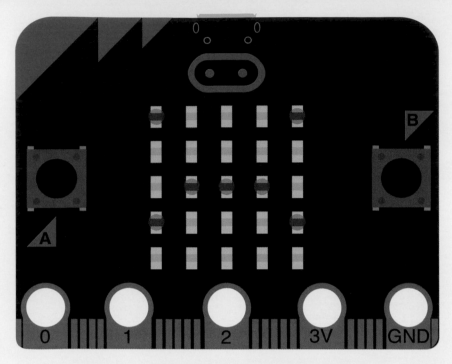

Animation
Exploration

Learn your way around the micro:bit by creating moving displays of cutting scissors and an exploding firework

Written and illustrated by Eric Hagan

ERIC HAGAN is an interactive and kinetic artist, and an assistant professor of sculpture at SUNY Old Westbury.

Make:
Easy micro:bit Projects

This skill builder is excerpted from Make: Easy micro:bit Projects — get the book and kit at makershed.com.

THE BBC MICRO:BIT IS A POWERFUL MINIATURE MICROCONTROLLER THAT CAN BE USED AS THE BASIS OF MANY EXCITING ELECTRONICS PROJECTS.

MEET THE MICRO:BIT

The front of the micro:bit (Figure **A**) contains a **5×5 LED grid**, two **programmable pushbuttons**, three labeled **digital or analog input/output rings**, and connection points for **3 volt power** and **ground**.

The back of the micro:bit (Figure **B**) contains the **micro-USB port**, a **reset button**, the **battery connector**, the **Bluetooth antenna**, the **processor**, the **compass**, the **accelerometer**, and **20 edge pin connectors**.

While there are several options for coding the micro:bit, we're using the **JavaScript Blocks editor** powered by **MakeCode** (Figure **C**). The editor is free at makecode. microbit.org, and it includes a "virtual micro:bit" so you can start learning to code even if you don't have a micro:bit yet.

When you create a new code project, the code blocks for **on start** and **forever** are automatically included (Figure **D**).

ON START

Code blocks attached to **on start** run exactly once when the micro:bit is turned on or reset.

FOREVER

Code blocks attached to **forever** will repeat over and over until the micro:bit is powered down or reset.

PROJECT: ANIMATED DISPLAYS

We are going to make animated LED "GIFs" of chopping scissors (Figure **E**) and an exploding firework (Figure **F**) while exploring *inputs* using the onboard buttons. We'll also learn how to use a few basic programming concepts like *variables* and *if/else conditional statements*.

CODING YOUR ANIMATIONS

1. Together the **show icon** and **show leds** code blocks (Figure **G**) will help us set which LEDs to turn on. Go to the Basic menu and drag one of each to the coding area. Both will appear gray, like in Figure

A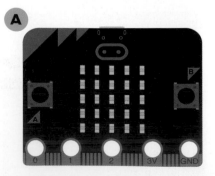

B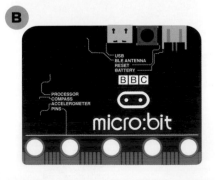

JAVASCRIPT BLOCKS EDITOR IN YOUR WEB BROWSER

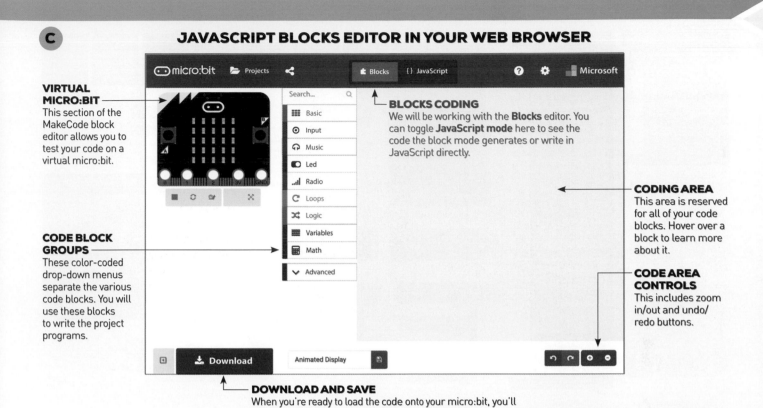

VIRTUAL MICRO:BIT
This section of the MakeCode block editor allows you to test your code on a virtual micro:bit.

CODE BLOCK GROUPS
These color-coded drop-down menus separate the various code blocks. You will use these blocks to write the project programs.

BLOCKS CODING
We will be working with the **Blocks** editor. You can toggle **JavaScript mode** here to see the code the block mode generates or write in JavaScript directly.

CODING AREA
This area is reserved for all of your code blocks. Hover over a block to learn more about it.

CODE AREA CONTROLS
This includes zoom in/out and undo/redo buttons.

DOWNLOAD AND SAVE
When you're ready to load the code onto your micro:bit, you'll give your project a name and click this Download button.

H (they need to be attached to an event in order to be enabled).

2. In the **show icon** block, click on the triangle to open the full menu. You'll start with the **scissors**, located in the lower right hand corner (Figure **H**).

3. Now attach the **show leds** block after the **show icon** block and drag them to the **forever** block (Figure **I**).

4. On the **show leds** block, you can just click on the individual LED rectangles to toggle them on. I have drawn the desired light pattern in Figure **J**, but basically you're just going to flatten the blades of the scissor into a straight line. Now our scissors are animated! You should see the scissors chopping on the LEDs of the virtual micro:bit.

5. Next you're going to add button presses. Go into the **Input** menu and select the **on button A pressed** code block and drag it into the code area (Figure **K**).

6. Detach the **show icon** and **show leds** from the **forever** block and attach them to the **on button A pressed**. Grab a second **on button** code block, click the triangle drop-down, and select the B button (Figure **L**).

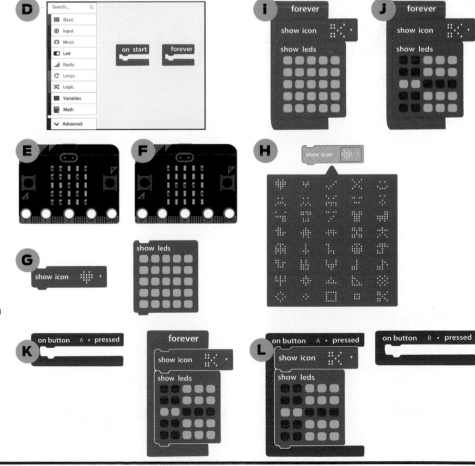

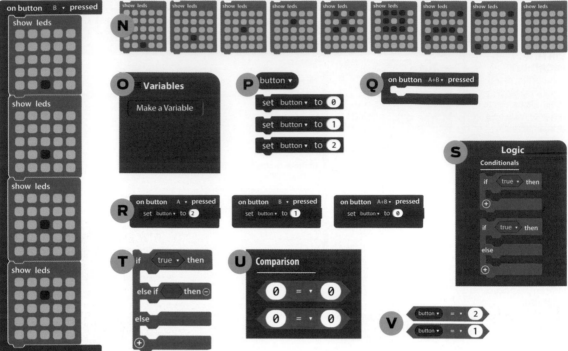

7. Under the **on button B**, you can copy and paste multiple **show leds**. Each of these blocks acts like a single frame of animation. Shown in Figure **M** are the first 4 frames for our firework animation.

8. Our firework takes nine total frames, which includes one blank frame with no lights on at the end (Figure **N**). Stack the blocks in order under the **on button B pressed** as in Figure M. Click button B on your virtual micro:bit to see the animation.

NOTE: Feel free to add as many frames as you want to this or any animation!

ADDING VARIABLES

You're going to add one more step to ensure that each animation plays continuously once the corresponding button is pressed. To accomplish this, you'll make a **variable**, and use a **conditional if/then statement**. You'll also add an **on button pressed** input block to make it so that pressing both the **A** and **B** buttons will turn the animations off.

1. First enter the Variable menu, and at the top select **Make a Variable** (Figure **O**). At the pop up window, name your variable **button**. Drag the **button** variable into the code area.

NOTE: A **variable** stores one piece of information with a name.

2. Next, drag three copies of the **set button to** code block into the code area. Again, these blocks will be disabled at this point and appear gray. Change the value of the second button variable to **1** and the third button variable to **2** by clicking inside the box and changing the number as shown in Figure **P**.

3. Grab one more **on button pressed** block from the input menu. You will set this one to turn off the animations when **A** and **B** are pressed (Figure **Q**).

4. Move the **show leds** animation blocks of the scissors and firework to the side, and then attach the **set button to 2** block to the **on button A pressed**, the **set button to 1** block to **on button B pressed**, and **set button to 0** on **button A+B pressed** (Figure **R**).

5. Now click on the Logic menu. You will use the **if then else** conditional block, the second from the top (Figure **S**).

NOTE: A **conditional statement** will perform different actions *if* a condition is met.

6. Click on the white **plus** icon in the lower left corner to add an additional **else if** to the conditional statement (Figure **T**).

7. Drag two comparison blocks to the code area (Figure **U**). These blocks will help you figure out which action to take with your conditional.

8. Make a copy of the **button** variable, then drag one button variable into the first section of the comparison block. Next, change the number to **2**. Do the same with the other comparison block, and change the number to **1** (Figure **V**).

NOTE: These **comparison statements** let us evaluate whether the listed statement is true.

9. Attach the first comparison block variable to the **if** box, then attach the scissor animation below the **if** statement (Figure **W**).

NOTE: With **conditionals**, it helps to say the results out loud in plain language. "If I press the A button, then the scissor animation will loop."

10. Attach the other comparison block to the **else if** section, and the firework animation to the corresponding **then**

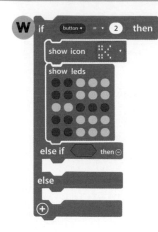

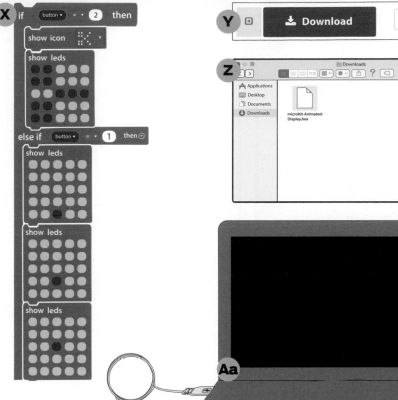

section (Figure **X**). This way, if we push the B button then our fireworks animation will play! Attach the entire piece to the **forever** block to enable it.

NOTE: You will leave the last **else** blank, since when you push both A and B the **button** variable is equal to zero, and nothing will be displayed, clearing the animations.

LOADING YOUR PROJECT ONTO THE MICRO:BIT

1. Now that the code is finished, input the project name "Animated Display" into the text box at the bottom of the code editor, then hit the Download button (Figure **Y**).

2. Your web browser will save a .hex file named *microbit-Animated-Display.hex* in the default download location on your computer. This is generally the *Downloads* folder (Figure **Z**). Open this folder.

3. Plug your micro:bit into your computer using the micro-USB cable (Figure **Aa**). Any available USB port should be fine. The micro:bit will show up on your computer as a USB drive with the name *MICROBIT*.

4. Drag the .hex file from your desktop to the *MICROBIT* drive (Figure **Bb**). The yellow power light on the back of your micro:bit

will flash. Now your code is uploaded and your animations will run!

NOTE: This is the same for both Mac and PC.

CAUTION: Do not unplug your micro:bit while you are uploading code as it could corrupt your program!

That's it! Now each of your animations will loop forever, until another button combination is pressed.

» When you push button A, this changes the value of the button variable to 2, and loops the scissor animation.

» If you push button B, this changes the button variable value to 1, which loops the fireworks animation.

» If you push both A and B at the same time, the animations will turn off. ◎

[+] You can find links to the code used in this project and the rest of the book at gitlab.com/MakerMediaBooks/easy_microbit_projects.

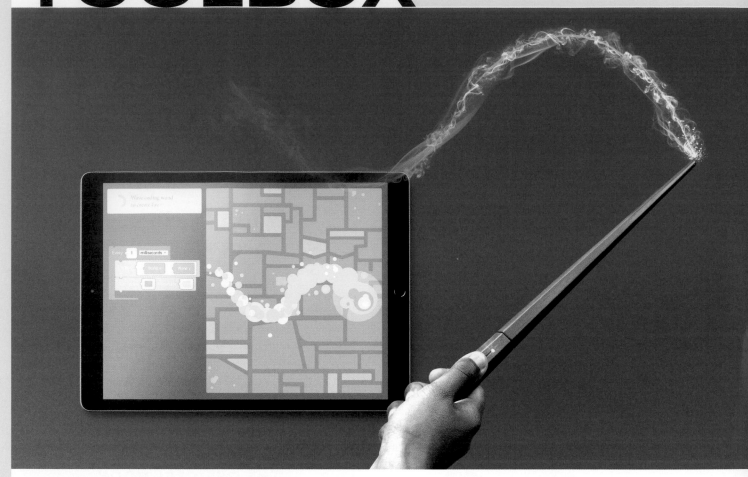

KANO HARRY POTTER WAND $99 kano.me

Known for their build-your-own computer for kids, Kano has conjured up more educational fun with a Harry Potter-themed wand. This coding kit lets you virtually swish and flick your way through the wizarding world while learning about logic, loops, and variables.

Assembly is quick and easy: slot the included batteries, insert the button, place the PCB, and snap on the cover. Doing it yourself reveals the magic: The wand connects to the app over Bluetooth and has a codeable vibration motor, light, and three sensors — a gyroscope, accelerometer, and magnetometer, which track its movements in the three-dimensional space.

The wand wasn't as responsive as my 10-year-old son and 8-year-old daughter would've liked, but that just made their first attempts at levitation with *Wingardium Leviosa* as ineffective as those of the Hogwarts students. Once they got the hang of it, they zipped through several of the 70+ coding challenges, such as directing flames and launching fireworks, leveling up to earn medals, rewards, and surprises, then experimenting beyond the directed steps. They traded off coding and casting of spells, and the results were often met with cheers, gasps, and comments like, "*Why* does it have to be *Snape*?!"

The verdict? Both deemed it "awesome," with my daughter exclaiming, "I feel like I'm Hermione!" —*Laurie Barton*

IPHONE SCREEN REPLACEMENT KIT

$24 (iPhone SE; prices vary for other models) amazon.com

Walking home after a dinner with friends in Las Vegas, I did the thing that all Vegas tourists do: pulled my phone out of my pocket to take a photo of one of the many oversized, overly lit spectacles along The Strip.

This time, however, my phone caught the corner of my pocket, cartwheeling out of my hand and through the air, then finally landing face-down with a sickening thud. I knew it before I even picked it up — the screen was shattered.

Apple wanted nearly $150 to replace the screen, but for that price I can buy a refurbished device, so I decided to first try my hand at DIY replacement. The $24 kit I found came complete with digitizer, tiny screwdrivers (including pentalobe), a magnetic mat for keeping parts organized, and more — everything I needed to complete the surgery.

I began the procedure itself at midnight, after a lengthy phone backup. The instruction book included small photos that left me crossing my fingers at times, and I'll admit that once or twice along the way I thought I had certainly ruined everything for good. But an hour later, after pushing the screen back in place and pressing the power button, the Apple logo blazed anew and I had my phone back in action once again, for a fraction of the price of repair or replacement. —*Mike Senese*

MILWAUKEE M12 FUEL COMPACT CUTOFF TOOL

$129 milwaukeetool.com/Innovations/M12

If you work with metal, you certainly use a heavy angle grinder nearly every day. For other jobs, you may need to use a small rotary tool. What if you need something in the middle, perhaps with an ergonomic handle and battery pack for portable use?

Milwaukee's M12 Cutoff Tool fits that space. It weighs in at just under 2.5 pounds with battery and cutoff wheel installed, and has a handle designed so that its pistol-style grip naturally aligns the cutting wheel with the user's forearm.

I got to check out one of these recently. Build quality appears to be very good, and the device made quick work of the scrap metal and plastic that I placed in front of it. One would think it would hold up quite well as a portable device to fill in the gap between large and small cutoff tools.
—*Jeremy Cook*

CANNIBBLE CUTTING DEVICE

$99 canibbletools.com/products/the-canibble-tool

When cutting sheet goods like thin aluminum or steel, there is always the struggle of getting the shears to do what you want while wrestling with the material. Due to the basic physics of shears, they tend to bend and warp the metal while the extra bits of your material are just getting in the way.

CanNibble approaches this challenge with a mechanical solution that takes tiny bites or nibbles from the material. It is an adapter that plugs into your drill, and is surprisingly effective. My tests on aluminum sheet and ABS sheet yielded super clean cuts with none of the creases I get when using sheers. The only downside is the pile of little "nibbles" left behind, and that's hardly a problem at all in a workshop regularly filled with metal shavings and sawdust. —*Caleb Kraft*

SHOW&TELL

Get inspired by some of our favorite submissions to Make: community

If you'd like to see your project in a future issue of *Make:* magazine submit your work to makershare.com/missions/mission-make!

1 Never worry about over- or under-watering your plants again! **Jonathan Pereira** developed Sprout, an automatic plant watering system. A moisture sensor monitors the soil and will give your plants a drink when the soil gets too dry. Pereira estimates that Sprout's 17fl oz tank can last as long as a month before needing a refill. makershare.com/projects/sprout-modern-indoor-self-watering-planter

2 Infinity mirrors are stunning to look at, with seemingly endless lights traveling to infinity. **Hal Breidenbach** took his version a step further by adding colored lights and turning it into a clock as a present for his grandson. makershare.com/projects/infinity-mirror-clock-light-show

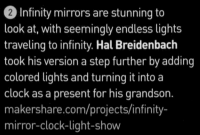

3 **Jacob Douglas** rigged up a simple Raspberry Pi and relay-controlled system to automatically turn a pair of powered speakers on and off in sync with his wireless music stream. The key was to add a delay to powering off so that the speakers wouldn't shut down every time the audio stream hiccupped. makershare.com/projects/simple-smart-speakers

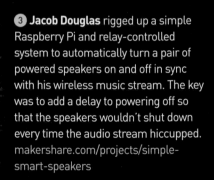

4 Along the same line as Sprout above, APEX is **Geoffrey McIntyre's** plant monitoring system that adds a little personality to your plant's needs. Using a moisture sensor, an Arduino Uno, and an 8×8 LED matrix, your plant will smile contentedly when recently watered and frown you into shame when its soil has become too dry. makershare.com/projects/artificial-plant-emotion-expressor-apex

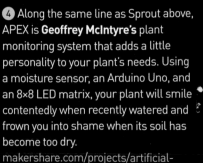

[+] Read about our Editors' Choice, the Sound Blink DIY portable speaker by Adhith Mathradikkal, on page 70.